MW00353268

RAILS AROUND THE WORLD

RAILS AROUND THE WORLD

Two Centuries of Trains and Locomotives

BRIAN SOLOMON

Contents

Introduction

In this book, I make historical examinations of a select group of significant locomotives and self-propelled trains. These were chosen for a variety of reasons, including novel design, significant or exceptional performance or longevity, wide geographic territory and technological influence, or milestone status. These locomotives include those that operated on countless railways around the world and are among those that changed railroading. Some types will be universal, such as the 4-4-0 American-type steam locomotive, which was the dominant North American locomotive in the nineteenth century and was built in large numbers for railways around the world. Others are more obscure, yet still remarkable and significant, and represent part of the greater story of world railways. In terms of numbers, Ireland's 121 class locomotives may seem almost insignificant, yet the type represents the first export of a General Motors diesel-electric model from its LaGrange, Illinois, factory to western Europe, and demonstrates the superiority of mid-twentieth-century American diesel technology over that of many traditional locomotive European manufacturers. For Ireland it was the most significant locomotive of the twentieth century.

Throughout the book, I have connected sections with common threads. I hope the largely chronological organization will help the reader recognize the significance of various locomotives and how they influenced subsequent development around the world.

In the nineteenth century, American and British builders helped establish technological precedents for railway propulsion everywhere. Although it doesn't get its own section, Britain's famous *Rocket* of 1829, built by Robert Stephenson, was unquestionably the most significant engine of all time. It set the pattern for most of the reciprocating steam locomotives built for mainline use by establishing the three basic design principles of a multi-tubular, fire-tube-type boiler; forced draft from cylinder exhaust; and the direct connection between the cylinders and driving wheels. This proved vastly superior to all previous locomotive designs and represented the granddaddy of world locomotive design until the advent of electric and later diesel motive power.

I've looked beyond the machinery to discuss the designers themselves, the hurdles they faced, and their ingenious solutions. I've tried to sample a wide range of types: steam, diesel and electric, passenger engines and freight, large locomotives and smaller ones too. My selection is intended to include railway locomotives used by many railways around the world, but it is not intended to be comprehensive. Rather than a mere listing of significant types, I wanted to tell the story of a select group of interesting and significant machines. There are many superlative locomotives that are not featured here but could have been. Perhaps those engines will be prominently featured in a future volume.

Operator controls for a high-speed German diesel-electric railcar like those assigned to *Flying Hamburger* services. Hundert Jahre Deutsche Eisenbahnen, Solomon collection

American 4-4-0

The American type, defined by its 4-4-0 wheel arrangement, was the dominant locomotive on mid-nineteenth-century North American railroads and one of the most enduring icons of American railroading. The wheel arrangement was introduced in 1836 by Philadelphia locomotive manufacturer Henry R. Campbell, who patented and constructed the first known 4-4-0 locomotive for the Philadelphia, Germantown & Norristown. A year later, builders Eastwick and Harrison significantly improved the type by introducing an equalization lever that provided the engine a three-point suspension. This suspension in combination with four driving wheels and a four-wheel leading truck made the 4-4-0 arrangement ideally suited for America's lightly built track structure. The American type was widely built from 1850 until the 1880s, after which new wheel arrangements facilitated larger locomotives, usurping the American's dominance. Yet many railroads continued to order new and larger 4-4-0s into the twentieth century. Despite obsolescence, some examples survived until the end of steam in the United States in the late 1950s and in Canada into the early 1960s.

The best remembered were the elaborately decorated and colorfully painted Victorian gems, dating from the early decades of commercial railroading, when most railroad locomotives were wood burners. Elaborate smokestack designs were employed to minimize damage from wood sparks, including conical balloon and diamond-shaped stacks, prominent characteristics of many mid-nineteenth-century engines. After the American Civil War, railroads gradually shifted to coal. Coal offered higher burn value and enabled construction of ever more powerful engines while increasing the production of the locomotive fireman, but coal soot was dirty. The coal era contributed to the end of the colorfully ornamental era for American engines and introduced more austere appearances, as many American railroads began painting their locomotives black.

↙ Nevada's Virginia & Truckee No. 12 was built by Baldwin in 1873 and has been restored to its early 1900 appearance for display at the California State Railroad Museum in Sacramento. *Brian Solomon*

↗ Matthias Forney featured this classic American in his book *Catechism of the Locomotive*, published in 1873 at a time when the 4-4-0 was the dominant locomotive in North America. This Rogers-built engine features the "Wagon top" style of boiler known for its tapered profile. *Catechism of the Locomotive, Solomon collection*

→ Wood-burning locomotives required elaborate smokestacks with screens and other spark arrestors to minimize the likelihood of line-side fires. A young girl observes this Plant System 4-4-0 simmering with a short passenger train. The late-nineteenth-century Plant System was a Georgia-based component of the Atlantic Coast Line. *Solomon collection*

EIGHT-WHEELED "AMERICAN" LOCOMOTIVE,
BY THE GRANT LOCOMOTIVE WORKS, PATERSON, N. J.
Scale, ⅛ in.=1 ft.

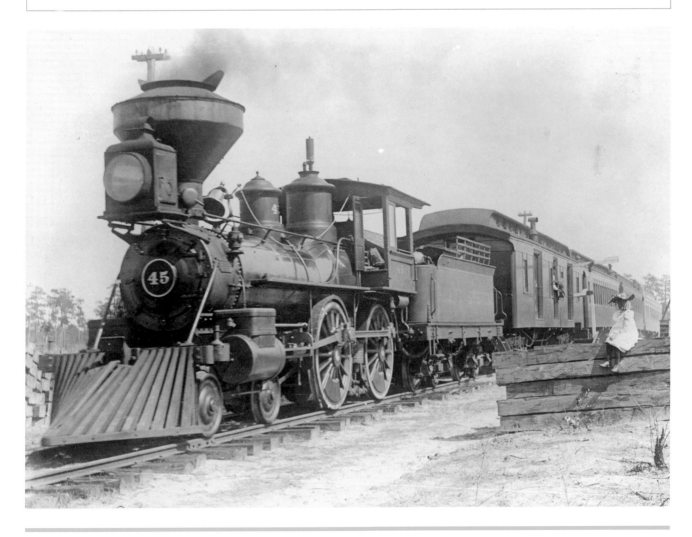

← Boston & Providence's *Daniel Nason* is a rare surviving example of a locomotive with cylinders located inside the smokebox (and thus concealed from view). It was designed by B&P's George S. Griggs and built in 1858. Today it is displayed at the Museum of Transport in Kirkwood, Missouri. *Solomon collection*

↙ Some railroads continued to operate 4-4-0s into the mid-twentieth century because these engines were comparatively inexpensive to operate and well suited to lightly constructed branch lines. Soo Line 4-4-0 No. 36 works a local passenger train at Superior, Wisconsin, on July 9, 1940. *Solomon collection*

↗ New York Central & Hudson River Railroad 871 is an example of a highly refined 4-4-0 built by the Schenectady Locomotive Works in 1890 for express passenger service. In its heyday, it would have worked the most famous "varnish" (as deluxe passenger trains were known) along the shore of the Hudson River. It was scrapped at the end of 1924. *Solomon collection*

The 4-4-0 was prized for its versatility, and certainly in the early years, many engines were built for general service. However, as railroads began operating heavier freights and faster passenger trains, more customized locomotive designs prevailed. Engines with taller drive wheels were designed for speed, while shorter drive wheels were better suited for starting and moving heavy freights. During the American's predominant period, the average size of engines with this wheel arrangement grew enormously. Where the early 4-4-0s weighed typically 12 to 15 tons, by the late nineteenth century, coal-burning 4-4-0s weighed between 40 and 45 tons, dwarfing the tiny wood-burners of the early period.

Among the most famous 4-4-0s were Western & Atlantic's *General*, know for its role in the Civil War's "Great Train Chase," today preserved at the Southern Museum of Civil War and Locomotive History in Kennesaw, Georgia; Central Pacific's *Jupiter* and Union Pacific's 119, which met face-to-face on May 10, 1869, to mark completion of the first Transcontinental Railroad at Promontory, Utah (where today replicas of these famous engines routinely recreate the historic event for visitors); and New York Central & Hudson River Railroad's 999, which on May 10, 1892, raced the *Empire State Express* toward Buffalo and was reported to have hit 112.5 mph (181 km/h)!

Camelback

During the nineteenth century, anthracite coal was sold as the preferred home heating fuel. Eastern Pennsylvania coal railroads thrived by moving growing tides of anthracite from mines to market. These railroads were rich with supplies of anthracite waste, but waste couldn't be easily burned in a conventional locomotive firebox. In 1877, the Philadelphia & Reading, the largest and most prosperous anthracite hauler, solved this problem when its general manager, John E. Wootten, developed a new type of locomotive using a wide, shallow firebox featuring a broad grate that enabled adequate anthracite combustion. Wootten's firebox was so big that it left limited space for the crew, requiring a compromised cab arrangement, so the engineer's cab was situated straddling the boiler *ahead* of the firebox, while a small platform for the fireman was located *behind* the firebox. Locomotives using this awkward split-cab arrangement were known as a "Camelback," distinct from the much older arrangement built by pre–Civil War manufacturer Ross Winans that had been descriptively known as a "Winans Camel."

Between the late 1870s and World War I, the Camelback arrangement was widely adopted by American anthracite carriers, including Central Railroad of New Jersey, Erie Railroad, Delaware & Hudson, Lackawanna, Lehigh Valley, Ontario & Western, and, of course, Reading Company. These were built using a variety of wheel arrangements, from modestly proportioned 0-4-0 switchers to Erie's enormous 0-8-8-0 articulated Mallet compounds. They were developed for both freight and passenger work. Some railroads also employed Camelbacks to burn other grades of coal, but these were the exceptions to the rule.

By World War I, anthracite steam locomotive design had matured, enabling anthracite-burning engines with conventional cab arrangements, and there were significant advances in bituminous-burning engine designs that effectively caused the Camelback arrangement to lose favor.

Camelbacks were unpopular with crews because they were difficult and unpleasant to operate. Both fireman and engineer's positions were uncomfortable, and separating communication between them proved demanding. Yet on a few railroads, Camelbacks survived through World War II, decades after they were deemed obsolete. Central Railroad of New Jersey was the last to operate them, working in commuter service from Jersey City until the early 1950s. A handful of the curious Camelbacks have been preserved, including examples from Reading Company, CNJ, and Lackawanna.

→ In 1906, an engine crew poses with Reading Company's Atlantic City Railroad 4-4-0 No. 23. This locomotive was built with very tall drive wheels for fast running on express passenger trains. Notice the engineer perched in the cab. *J. E. Carney, Solomon collection*

↘ The largest Camelbacks were Erie Railroad's three massive 0-8-8-0s Mallet compounds built by Alco in 1907. These were based at Susquehanna, Pennsylvania, for helper service on heavy eastbound freights crossing Gulf Summit. *Photographer unknown, Solomon collection*

↖ Intensively operated eastern Pennsylvania anthracite lines originated both the 2-8-0 wheel arrangement and Camelback cab arrangement, two significant innovations exhibited on Reading Company 2-8-0 Consolidation 1575, built as an anthracite burner for heavy freight service. *Photographer unknown, Solomon collection*

← The dual-cab Camelback was invented by Philadelphia & Reading's John E. Wootten to allow for a broad, shallow firebox. P&R 4-4-0 349, built in 1886, was an early example of the Camelback arrangement that featured an ornately decorated engineer's cab. *Photographer unknown, Solomon collection*

↑ Camelback 4-6-2 Pacific types were very rare. This Lehigh Valley Pacific was built in 1906 and is pictured at Allentown in the 1920s. *Chas. E. Fisher, Solomon collection*

Class J15 0-6-0

Great Southern & Western Railway's (GS&WR) Class 101 0-6-0 (known after the creation of Great Southern Railways in 1924 as Class J15) was an engineering classic and truly a "go anywhere, do anything" workhorse locomotive. The J15's long careers on Irish railways reflected reliability, versatility, and ease of maintenance and operation. Born of the nineteenth century, this type was built in greater numbers than any other Irish locomotive, surviving in traffic until supplanted by diesels in the 1950s and early 1960s.

Great Southern & Western Railway was the most important and the most extensive Irish railway. It was focused on Dublin and, in accordance with Irish practice, built its lines using 5-foot-3-inch (160 cm) gauge tracks. In 1864, Dublin-born and -educated Alexander McDonnell, after having studied railway operations in England and Europe, assumed the head of GS&WR's locomotive department. Soon after assuming his position, McDonnell

worked with Beyer, Peacock & Company of Manchester, England, to design a new goods (freight) locomotive; this emerged as the basic, yet powerful, 0-6-0 engine that originated the 101-class pattern.

The first prototypes were rebuilt from older locomotives at GS&WR's Inchicore Works in Dublin. Later locomotives were variously built by Beyer, Peacock, Atlas Works of Sharp, Steward & Company, and at Inchicore. This resulted in a total of 119 machines of the class by 1903, plus a pair of similar engines for the Dublin & Belfast Junction Railway.

Although numerous changes were introduced during the J15's long build period and service lives, perhaps the most successful change to the basic design was the introduction of superheating in the early twentieth century. Not all the J15s were modified, and some remained as traditional "saturated" engines until retirement in the 1960s.

The Railway Preservation Society of Ireland (RPSI) was founded in 1964 and has operated

↙ It was a misty afternoon in County Kerry on May 6, 2006, when J15 186 hauled a Railway Preservation Society of Ireland excursion across the famous Quagmire Viaduct en route to Killarney. *Brian Solomon*

→ Railway Preservation Society of Ireland J15 186 marches west with an excursion on the Sligo Line at Hill of Down in 2007. *Brian Solomon*

historic mainline excursions north and south of the Irish Republic/Northern Ireland border ever since. Its fleet of preserved Irish steam includes two examples of the J15: engine 184 is a traditional saturated engine that was built at Inchicore in 1880, while 186 was built by Sharp, Stewart & Company in 1879—an example of a superheated engine rebuilt in the twentieth century. Both engines were regular performers in the 1960s and 1970s, but damage to 184 has limited it to a static display in recent years. Engine 186 was restored to service in 2004 for RPSI's 40th anniversary and worked excursions for a decade before being taken out of traffic in 2014. Both are stored at RPSI's Whitehead shops in Northern Ireland.

⬉ Locomotive 186 was restored to approximate its twentieth-century appearance in minimalist charcoal gray, as seen here. The J15 was designed as a utilitarian functional machine and was not afforded elaborate adornment.
Brian Solomon

→ RPSI 186 was enveloped in its own steam during a water stop at Portlaoise in 2006. This was among the J15s that were rebuilt with superheaters to increase the power of the locomotive. Sister locomotive 184 survived as a "saturated" engine (not superheated) until the end of regular steam service in Ireland.
Brian Solomon

⬊ The shrill cry of 186's whistle shatters the evening silence at Farrenfore, County Kerry. The Railway Preservation Society of Ireland keeps the spirit of steam alive with its small fleet of active locomotives.
Brian Solomon

Consolidation 2-8-0

Consolidation is another term for merger. In 1865, Alexander Mitchell, master mechanic of Pennsylvania's anthracite coal hauler Lehigh and Mahanoy Railroad, needed a more powerful and agile heavy freight locomotive and expanded the 0-8-0 into a 2-8-0 by adding a radial two-wheel pony truck to help the big engine better negotiate curves. He contracted with Philadelphia's Baldwin to build his nonstandard design, which the prolific locomotive builder accepted only reluctantly, as it did not approve of Mitchell's advancement. During construction, the Lehigh and Mahanoy was merged into the Lehigh Valley, which led to a promotion for Mitchell and to Lehigh Valley inheriting the new locomotive; as a celebration of the joining of the two lines, the 2-8-0 wheel arrangement was named the Consolidation type.

As railroads grew, "big" was relative, and what began as an unusually large locomotive gradually emerged as a standard type. By the turn of the nineteenth century, the 2-8-0 had become the most common type of road freight engine in North America. The type was

continuously built for a half century, leading to an estimated 23,000 Consolidations constructed domestically for American railroads and an additional 10,000 built for other countries. Globally, the 2-8-0 was adopted and adapted for myriad applications, making it one of the most common types worldwide.

← Union Pacific was an early buyer of the 2-8-0 Consolidation and sampled one from Baldwin in 1868. Pictured here is Taunton Locomotive–built U-class 268, constructed in 1883. It was characteristic of early UP 2-8-0s, with its diamond stack, polished boiler plate, and ornate steam and sand domes. In its day, this would have been considered a *huge* locomotive. *Solomon collection*

← American manufacturers built 2-8-0s for export to railways around the world. This Alco 2-8-0 was built by the American Locomotive Company in the early twentieth century for the Central Portugalete railway. *Alco photograph, Solomon collection*

↙ Southern Pacific Class C-9 2-8-0 No. 828 was a classic workhorse freight locomotive of the early twentieth century. Some members of this class worked for a half century before being replaced by diesels in the 1950s. *C. W. Witbeck photo, Solomon collection*

← Denver & Rio Grande bought this 3-foot gauge 2-8-0 from Baldwin in 1881. It has been preserved at the Colorado Railroad Museum in Golden and is a rare example of a serviceable nineteenth-century American steam locomotive. *Brian Solomon*

↙ Maine Central 501 was built by Alco for heavy freight service on the railroad's Mountain Division over New Hampshire's Crawford Notch. Once part of Nelson Blount's Steamtown collection, it is preserved today at the Conway Scenic Railroad in North Conway, New Hampshire, and may be restored to service for operation on home rails. *Brian Solomon*

↙ Western Maryland Scenic 2-8-0 Consolidation No. 734 was originally a Lake Superior & Ishpeming locomotive. It is seen working through the famous Helmstetter's Curve on the climb to Frostburg, Maryland. This was a late-era machine with a large boiler that was equal or greater in power to many moderate-size 2-8-2s. *Brian Solomon*

Stanier 8F Consolidation

In the twentieth century, the 2-8-0 was refined on British and European railways and adapted as a standard freight engine around World War I, remaining standard through the end of steam. One of the best British 2-8-0 designs was the work of William Stanier (1876–1965), a prodigy of Britain's Great Western Railway. He was apprenticed to the company's Swindon Works as a teenager and advanced through the ranks under GWR's chief mechanical officers George J. Churchward and C. B. Collett. In 1932, Stanier was recruited to be London Midland Scotland's Chief Mechanical Officer (CMO), building his reputation by focusing GWR's most effective locomotive design practices into new standardized types for LMS. Among his best known, and by far his most numerous engine design, was his mass-produced Class 8F 2-8-0 introduced in 1935. This totaled 852 locomotives, including 133 engines constructed during World War II for the War Department. Many were sent to the Mideast, including Egypt, Iran, and Turkey, among other places, with some engines working for decades after the war. Notably, some Turkish examples worked revenue trains into the 1980s. The 8F was prized for its relatively light axle load and high tractive effort, enabling it to serve on many routes and yet handle some relatively heavy trains. It was known for supplying ample amounts of steaming characteristics with economical use of fuel and water and for being light on maintenance.

↙ Restored British Railways 2-8-0, Class 8F 48431 simmers on the Keighley & Worth Valley, where it was entertaining visitors on this popular heritage railway. *Brian Solomon*

→ Stanier's Class 8F 2-8-0 was among the most widely built steam locomotives in England. Engine 48151 leads a mainline passenger excursion at Crewe, England, in 2003. *Brian Solomon*

↘ The 8F is a classic British steam locomotive that exhibits handsome, well-balanced design that served for decades under demanding conditions. *Brian Solomon*

Fairlies

In the 1860s and 1870s, Robert F. Fairlie was among the leading British promoters of narrow gauge railways. Narrower tracks, a tighter loading gauge, and sharper curves required less intensive infrastructure, which made narrow gauge railways cheaper to construct in mountainous areas and in other places where standard gauge railways would have been deemed as too costly. In the 1860s, Fairlie invented, patented, and built a distinctive double-ended twin-boiler locomotive. Instead of a rigid wheelbase with driving wheels fixed to the frame of the engine, his novel locomotive employed twin pairs of swiveling trucks, one beneath each boiler, with a central cab situated between the boilers. Because the total weight of the engine was placed on its driving wheels at all times, Fairlie claimed that his arrangement would match the power of a standard gauge engine while enabling it to negotiate very tight curves without loss of adhesion.

Fairlie was a visionary: he named his first locomotive *Progress*. This was built for the obscure Neath & Beacon Railway in 1865, but his type caught the attention of the railway world four years later, in 1869, when a 2-foot gauge Fairlie named *Little Wonder* was built for the Welsh slate-hauling Festiniog Railway. Festiniog was among the leading users of Fairlies in Britain, employing them for more than seventy years. Fairlie, meanwhile, went on to sell larger examples of his patented engine to railways around the world, largely in Central and South America, Africa, and Australia. America's pioneer 3-foot gauge Denver & Rio Grande bought a lone Fairlie named *Mountaineer*, built in England by the Vulcan Foundry and delivered to Colorado in June 1873. This was typical of the improved double-ended Fairlies and featured twin fireboxes to overcome problems with draft. It was largely used in helper service over Colorado's Veta Pass crossing. Mexico also bought Fairlies.

The Fairlie type, although a relatively obscure engine in terms of numbers built, is of great interest to locomotive historians because its swiveling driving wheels anticipated the now-standard arrangement of powered twin-axle trucks used by electric railcars and locomotives and, later, diesel-electric locomotives around the world.

A variation of the Fairlie was a single boiler, single truck, swiveling locomotive design devised by William Mason, an innovative New England builder. Although Mason called these "Mason-Fairlies," they were more commonly known as "Mason Bogie" engines. Roughly ninety Mason Bogies were built for American narrow gauge lines, including Boston, Revere Beach & Lynn, and Denver, South Park & Pacific. Although another relatively rare type, the Mason Bogie engines were a pioneer application of Walchaerts outside valve gear. In the twentieth century, this gear emerged as the new standard valve gear arrangement in North America for most new locomotive types, and it aided in the development of much larger and more powerful steam locomotive designs.

Today, the preserved Ffestiniog Railway (two *ff*s is now the preferred Welsh spelling) continues to entertain visitors with double-ended Fairlie steam locomotives. Its *Merddin Emrys* dates to 1879, while modern replicas also regularly work on this scenic narrow gauge line, which connects its namesake Welsh mining town with Porthmadog, where it makes a connection with the recently reconstructed Welsh Highland Line.

→ On the preserved Ffestiniog Railway, the tracks make a complete loop or helix to gain elevation in the mountains. This alignment deviates from the historic narrow gauge railway to avoid a dam. *Brian Solomon*

↘ Ffestiniog's *Earl of Merioneth* was built in 1979, more than a century after Fairlie's first successful double-ended narrow gauge engines. *Brian Solomon*

↓↓ The double-ended Fairlie has been a fixture of the Welsh mountains around Ffestiniog for generations. *Brian Solomon*

Forney 0-4-4T Tanks

atthias N. Forney was a Victorian-era polymath: a practicing journalist, a successful businessman, and a locomotive designer among America's best-known railway engineers of the time. He was editor and part owner of the *Railway Gazette*, a leading railway trade publication of the time. His *Catechism of the Locomotive* was a standard locomotive educational text. Forney had learned the locomotive business from Ross Winans, one of America's pioneer engine manufacturers. Rather than advancing and adapting British practices, Winans designed distinctive American locomotives, and Forney learned from Winans's innovative approach.

In the 1870s, Forney applied his genius to the design of a compact, bi-directional tank engine for suburban passenger services. An engine built for short runs, rapid acceleration, and quick turnaround time at terminals required an unusual configuration as compared to a road locomotive hauling long-distance trains.

Conventional steam locomotive design required the engine to carry fuel and water in a separate, semipermanently coupled tender, which wasn't conducive to bi-directional operation. Switching service tank engines suffered from low fuel and water capacity; as their fuel and water supplies were depleted, adhesive weight on driving wheels was lowered, reducing pulling power. Furthermore, typical tank engines were impaired by a short, rigid wheelbase without guide wheels, ill-suited for fast running.

Forney's solution was to use the previously untried 0-4-4T wheel arrangement on an extended rigid frame to accommodate a fixed tender. This placed full adhesive weight on driving wheels while avoiding operational disadvantages of a short wheelbase by matching the average wheelbase of a typical contemporary 4-4-0 road locomotive. Forney's trailing truck design below the tender section of his locomotive both minimized the undesirable effects of coal/water depletion on adhesion (a common problem for tenderless

↙ This illustration from Matthias Forney's book *Catechism of the Locomotive* displays the inventor's concept for the 0-4-4T as a bi-directional tank locomotive. *Catechism of the Locomotive, Solomon collection*

→ New York City's elevated lines employed fleets of small steam locomotives prior to electrification at the turn of the nineteenth century. This compact 0-4-4T Forney type worked the 6th Avenue Elevated line that ran from South Ferry to 53rd Street and 9th Avenue in Manhattan. *Richard Jay Solomon collection*

→ This diminutive 0-4-4T Forney was an 1877 product of Boston's Hinkley Locomotive Works, built for the short-lived 2-foot gauge Billerica & Bedford. The B&B folded after a year's operation, and 0-4-4T *Puck* was later sold to Maine's Sandy River & Rangeley Lakes Railroad. *Solomon collection*

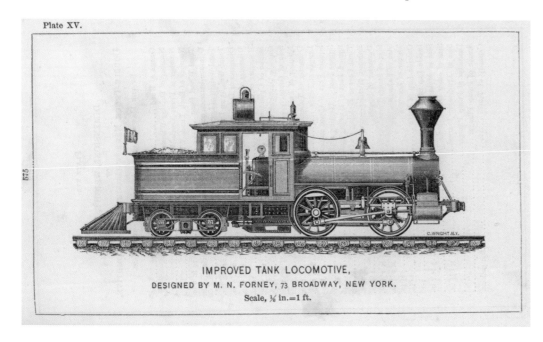

Plate XV.

IMPROVED TANK LOCOMOTIVE,
DESIGNED BY M. N. FORNEY, 73 BROADWAY, NEW YORK.
Scale, ¼ in.=1 ft.

tank locomotives) and allowed for a fully bi-directional locomotive well suited to both rapid acceleration and quick terminal turnaround without the need for a turntable.

Forney's 0-4-4T engines were known simply as "Forneys" and proved well adapted to steam-powered rapid transit. Among the most intensive operation of Forneys was on elevated railways in New York City and Chicago. Compared with road locomotives for mainline use, rapid transit Forneys were relatively small and lightweight but powerful, making them well suited to elevated railway operations. When the Els ultimately electrified operations, the cast-off Forneys found work as logging engines, where a short, flexible wheelbase, light axle weight, and bi-directional operation was desirable. A few of these made it to Alaska, where some survived derelict for decades after the logging railroad they powered had concluded operations. Some Forneys were

built for heavy railroad suburban services, notably Illinois Central's Chicago commuter runs. Ironically, while Forney himself was vocally opposed to development of narrow gauge lines, his 0-4-4T was adopted by a number of new slim gauge railroads, notably in Maine, where they survived in service longer than anywhere else.

← A variation on the Forney design was the 2-4-4T, which incorporated a leading pony truck for better tracking qualities at speed. Illinois Central operated large numbers of Forneys and other types of tank passenger locomotives on its intensive Chicago suburban service until it electrified these lines in the 1920s. *Solomon collection*

← Bridgton & Saco River No. 7 is a 2-foot gauge 2-4-4T built by Baldwin in 1913 and is among the active locomotives operated by Maine Narrow Gauge Railroad Company & Museum. It is seen here at Sheepscot, Maine, during an operational visit to the Wiscasset, Waterville & Farmington in January 2020. *Brian Solomon*

→ Over the last thirty years, a group of dedicated preservationists have restored a section of former Wiscasset, Waterville & Farmington 2-foot gauge line and former WW&F 0-4-4T Forney No. 9. On January 18, 2020, No. 9 made a spectacular display of steam near Alna, Maine. *Brian Solomon*

Compound Types

Compound steam locomotives were devised as a means of improving steam locomotive efficiency and operation. A conventional single-expansion steam engine suffers from inefficiency because considerable expansive power remains in pressurized steam when it is exhausted from the engine. Recognizing this flaw, locomotive designers aimed to improve efficiency, and thus conserve water and fuel consumption, by devising various schemes for double-expansion engines. Known as compounds, these designs take greater advantage of steam expansion by directing exhausted steam into a second set of cylinders before exhausting it into the atmosphere. British and European engine builders developed compound designs as early as the 1870s, while the earliest example of an American compound was an experimental built by Shepard Iron Works in Buffalo, New York, in the late 1860s. During the 1880s, numerous American railroads sought improved efficiency and adopted a variety of non-articulated, double-expansion locomotive arrangements using the exhaust from high-pressure cylinder(s) to power low-pressure cylinder(s). In America, non-articulated

compound engines were commercially built in large numbers from the late 1880s until about 1910, when they largely fell out of favor because the greater maintenance required to keep compound engines in good order often exceeded fuel savings; the adoption of superheating after 1906 allowed for significantly greater steam efficiency with less complex machinery. In Europe, compound designs remained popular to a limited extent until the 1950s.

In order to provide approximately equivalent output, the diameter of a low-pressure cylinder must be more than twice that of the engine's high-pressure cylinders, so compound engines employed a variety of high- and low-pressure arrangements. The two-cylinder, cross-compound featured a single high-pressure cylinder on one side of the engine with a low-pressure cylinder on the other. Because the low-pressure cylinder needed to provide equal power to the high-pressure cylinder, and it used a substantially larger diameter, the locomotive had an unbalanced asymmetrical appearance. While in theory the two cylinders supplied equal force, in practice the result was often uneven, leading to some crews calling the engines "slam-bangs." Many

Baldwin built ten four-cylinder balanced compound 4-4-2 Atlantics for America's Great Northern Railway in 1906. These were Class K-1 numbered 1700 to 1709. High-pressure inside cylinders were coupled to the first pair of drivers, while outside low-pressure cylinders power the rear pair of drivers. *Solomon collection*

← Colorado's Manitou & Pike's Peak Railway employed specially designed Vauclain compound steam locomotives on its steeply graded rack-railway line. These featured pairs of high-pressure and low-pressure cylinders on both sides of the engine. The high-pressure cylinders on the bottom exhausted steam directly into the larger low-pressure cylinders above to improve thermal efficiency. *J. William Vigrass*

← The articulated Mallet compound was introduced to North America in 1904 with Baltimore & Ohio's Alco-built 0-6-6-0 No. 2400. Similar to that pioneer was Kansas City Southern 0-6-6-0 700, one of twelve Alco Mallets built for the line in 1912. KCS was unusual because it used 0-6-6-0s to lead freights in road service. These locomotives were more commonly used as helpers or yard engines. *Photo courtesy of Harold Vollrath, Solomon collection*

were converted to the conventional arrangement after just a few years.

More successful were non-articulated four-cylinder compound arrangements. Best known in the United States was Baldwin's Vauclain type, named for its inventor, Samuel Vauclain, who patented this type in 1889. This type used matched pairs of high- and low-pressure cylinders one on top of the other on each side of the locomotive. Each pair powered a common cross-head connection and featured a cylindrical piston valve to provide steam admission to both cylinders that had a distinctive appearance, with a triple cylindrical cluster on each side of the engine. These were among the earliest commercially produced locomotives to make widespread use of the piston valve, offering superior valve and port arrangements necessitated by an unusually complex steam passage arrangement integral to the Vauclain design.

French locomotive designer Alfred G. DeGlehn designed a four-cylinder, four-crank balanced compound type that featured all four cylinders positioned in a horizontal row, which employed four crank points instead of two. The most common DeGlehn arrangement used two high-pressure cylinders located inside locomotive frames that powered lead driving wheels with a cranked axle; the low-pressure cylinders were located outside the frame in a more conventional position, with external rods connected to the outside crank pins of a second set of drive wheels. Baldwin's Samuel Vauclain refined a similar arrangement known as the Baldwin Balanced Compound.

Various arrangements of three-cylinder compound engines gained favor in Britain,

George T. Glover of Ireland's Great Northern Railway adapted the three-cylinder compound into a powerful, fast 4-4-0 express passenger engine in order to overcome the limitations of the conservative Irish weight restrictions. Each of his V-class compounds was named after birds of prey, numbered from 83 to 87: *Eagle, Falcon, Merlin, Peregrine*, and *Kestrel*. The Railway Preservation Society of Ireland preserved *Merlin*, seen here at Laytown along the Irish Sea. *Brian Solomon*

Ireland, and elsewhere in Europe. Typically, these involved a center cylinder powering a cranked axle. Although Reading Company built a few experimental three-cylinder compounds for express passenger service, this arrangement was not favored in North America.

The largest and most powerful compounds were American adaptations of the European articulated Mallet compound—a type that was, in effect, two engines articulated beneath a common boiler. This began in 1904 with an Alco-built 0-6-6-0 for Baltimore & Ohio and culminated with the massive Y6 class 2-8-8-2s built by Norfolk & Western at Roanoke, Virginia. The majority of American Mallets were built as 2-6-6-2s and 2-8-8-2s. Among the most peculiar Mallets were oil-burning, cab-forward types built by Baldwin for Southern Pacific; gargantuan, behemoth Triplex types with three sets of running gear built for Erie Railroad and Virginian; and bizarre, flexible boiler types constructed for Santa Fe.

↙ Manitou & Pike's Peak's Vauclain Compound No. 1 is displayed at the Colorado Railroad Museum in Golden. It is one of only a scarce few surviving examples of Baldwin's Vauclain compound design. *Brian Solomon*

→ The final examples of the Mallet compound were Norfolk & Western's massive Y-6 class 2-8-8-2s, built by the railroad's Roanoke Shops for coal service. On July 31, 1958, an N&W Y-6 works at the back of a loaded coal train at Blue Ridge, Virginia. *Richard Jay Solomon*

↘ In 1899, photographer I. Walter Moore photographed a brand-new Schenectady Locomotive Works–built cross-compound at his local station in Warren, Massachusetts. Two of these 4-6-0s were built for Boston & Albany and known as "slam-bangs" by crews because of their unbalanced piston action. *I. W. Moore photo, Robert A. Buck collection*

Shay Geared Types

The conventional reciprocating rod-driven steam locomotive faced limitations on industrial lines with unusually steep grades and lightly built undulating track. To work these lines, specialized steam locomotives were designed for slow-speed industrial service using a geared power train that allowed for high tractive effort on rough track. The first and most common of these was the Shay type, designed by Ephraim Shay in 1878 and built by the Lima Locomotive Works of Lima, Ohio. The Shay type was a curious example of an asymmetrical locomotive using groups of vertical cylinders positioned on the engineer's side of the locomotive to power a horizontal crank shaft that in turn used gearing to drive pivoting two-axle trucks at each end

of the locomotive—a wheel arrangement that anticipated the common diesel-electric locomotive invented decades later. This provided high adhesion at slow speed with necessary flexibility for minimal risk of derailment on poor track. Later the Shay was expanded into a three-truck version.

Shays were built well into the twentieth century, and the final example was also by far the largest: a huge three-truck Shay constructed in 1945 for Western Maryland. This engine, along with other Shays, is preserved on West Virginia's Cass Scenic Railroad, one of several tourist lines operating geared engines in excursion service.

In the steam era, geared locomotives were commonly used on lumber railways, well known for rough temporary trackage used to reach timber stands. These railways operated in northern New England, the Appalachians, the Upper Midwest, California, and the Pacific Northwest. New York Central employed specially built enclosed Shays that shrouded reciprocating parts for its New York City street trackage, where it was feared that a locomotive's reciprocating components might spook horses.

Among the less common varieties of geared locomotives were those built in Pennsylvania by the Climax Manufacturing Company of Corry, which used a pair of parallel, steeply angled cylinders on opposite sides of the boiler to drive the geared power train; the Heisler Locomotive Works in Erie, Pennsylvania, built engines featuring a pair of angled cylinders oriented at a 90-degree angle from each other and situated crosswise below the boiler.

↙ Many Shays featured a cast Lima number plate at the front of the engine. This engine is displayed near Lima, Ohio, where the Lima Locomotive Works built locomotives until the early 1950s. *Brian Solomon*

→ West Virginia's Ely Thomas Lumber Company was still a going concern on June 1, 1958, when its three-truck Shay No. 2 was pictured under steam on Kodachrome slide film. *Richard Jay Solomon*

↘ Preserved 3-foot gauge *Dixiana* is an example of a two-truck Shay. It was built in 1912 for a Tennessee lumber company and worked for several owners before finding a home at the Roaring Camp & Big Trees tourist line in Felton, California. *Brian Solomon*

↑ Pacific Coast No. 2 is among the active Shays at West Virginia's Cass Scenic Railroad. All the cylinders on a Shay are located on the right-hand side of the locomotive, which makes this a rare example of an asymmetrical steam locomotive design. *Adam Stuebgen*

↗ The last and by far the largest Shay was Western Maryland No. 6, built by Lima in 1945. This monster three-truck Shay is seen at work with two of its smaller sisters at West Virginia's Cass Scenic Railroad. *Adam Stuebgen*

→ New Hampshire's White Mountain Central operated by Clarks Trading Post in Lincoln owns several geared steam locomotives, including this two-truck Shay lettered for the historic Woodstock Lumber Company. *Brian Solomon*

Rio Grande Narrow Gauge Mikado 2-8-2

When new in 1903, Rio Grande's narrow gauge 2-8-2 Mikados were the "monsters" of the Rockies. Step back a few years: in the 1890s, Baldwin first developed the 2-8-2 wheel arrangement for export, and the type received its colorful moniker because Japanese Railways was an early buyer. At the time, Gilbert & Sullivan's opera *The Mikado* enjoyed great popularity in the United States, so the name stuck.

Among the first domestic applications for the Mikado was on Denver & Rio Grande, which bought fifteen 3-foot gauge Mikados from Baldwin in 1903. These were Class K27 engines, originally deemed monsters because of their relative size to earlier narrow gauge engines. Later, Rio Grande's K27s earned the popular moniker "Mudhens" because of their squat appearance and the propensity to stir up dust on dry days.

These unusual locomotives blended several evolutionary advancements that stood out from typical American narrow gauge engines. Notably, they used outside frames instead of the standard inside frame arrangement and featured outside counterweights and crankpins.

As built, DRG's K27s were Vauclain compounds (see compound engines on page 32) that used two sets of high- and low-pressure cylinders and were the only narrow gauge compounds built for Rio Grande. Later, these were rebuilt into simple locomotives. In their early days, Rio Grande's K27s were initially restricted to service on Marshall Pass. Later, the railroad upgraded the west slope of Cumbres Pass to enable the K27s to work as helper locomotives between Chama and the summit.

In 1923, Alco supplied Rio Grande with twelve 3-foot gauge 2-8-2s, known as Class K28s, that were nominally larger than the K27s and more powerful, being capable of working at speeds up to 40 mph (64.4 km/h), which made them well suited to passenger traffic. Two years later, Baldwin built ten more 2-8-2s, Class K36, even larger than the K28s. Finally, between 1928 and 1930, the railroad's Burnham Shops converted standard gauge 2-8-0s into powerful narrow gauge steam locomotives, Class K37. Operational examples of Rio Grande's narrow gauge Mikados survive on the Durango & Silverton and Cumbres & Toltec scenic railroads in Colorado and New Mexico.

↙ The Class K27 was the first and smallest of Rio Grande's narrow gauge Mikados. Fifteen 3-foot gauge K27 2-8-2 Mikados came from Baldwin in 1903. *J. R. Quinn photo, Solomon collection*

→ Rio Grande K36 487, fitted with a large plow, double heads with K27 453 over Cumbres Pass. Considering how relatively few late-era narrow gauge steam locomotives Rio Grande bought, it is remarkable how many of them have survived into modern times. *J. R. Quinn photo, Solomon collection*

↑↑ Cumbres & Toltec former Rio Grande K36 Mikado No. 484 works near Windy Point, New Mexico, on the climb toward Cumbres Pass, Colorado. *Brian Solomon*

↑ Among the peculiarities of Rio Grande's 3-foot gauge Mikados was their design employing outside frames, outside counterweights, and crankpins, in contrast to conventional North American practice, in which these components were concealed between the locomotive wheels. *Brian Solomon*

→ Rio Grande K27 463 leads a double-headed Cumbres & Toltec passenger train on its ascent of Cumbres Pass in September 1998. *Brian Solomon*

Doodlebugs and Wind-Splitters

Since the formative days, railway companies had experimented with small, self-propelled railcars for service on lightly traveled lines. In the nineteenth century, these were typically powered by small onboard steam engines. Later railcars employed novel and sometimes unsuccessful methods of propulsion. Commercial development of the self-propelled electric streetcar in the 1880s led to experiments with railcars powered by gasoline engines coupled to a generator supplying electricity to traction motors. William Patton was among the early pioneers, and in 1890, he built a very small gasoline-electric railcar using a compact 10-horsepower Van Duzen engine. Although Patton gave up building self-propelled railcars in 1893, his concepts were advanced by other manufacturers in the twentieth century, and gasoline-electric powered railcars emerged as a popular means for cost cutting on branch lines.

In 1904, Edward Henry Harriman, who controlled both Union Pacific and Southern Pacific, urged UP's motive power chief, William J. McKeen, to develop a practical railcar suitable for branch-line passenger service. Inspired by developments with internal combustion engines, McKeen worked with UP's engineers at the company's Omaha Shops and in March 1905 produced a lightweight railcar prototype with an aerodynamic design using a knife-edged front end and a rounded back intended to minimize wind resistance and improve adhesion (though it turned out that when the cars were turned around with the rounded end at front, resistance was reduced better than with the knife edge facing forward). The McKeens were powered by a gasoline engine with a mechanical transmission. In 1908, the McKeen Motor Car Company began production at UP's shops and for the next decade supplied roughly 150 cars to railroads in the United States, as well as Australia, Canada, Cuba, and Mexico. The majority of the cars were bought by Harriman lines, Union Pacific, Southern Pacific, and affiliated companies. The car's engine placement

This hand-tinted penny postcard photo features one of Union Pacific's many McKeen motor cars with an unpowered trailer at Greeley, Colorado, about 1910. Most of UP's McKeen cars were powered by a 200-horsepower gasoline engine. *Solomon collection*

Arizona Eastern car No. 3 was a 35-ton, 70-foot-long McKeen motor car built in 1911. Notice the oversized steam locomotive headlight and "cow catcher" pilot at the front of the car. In the late 1920s, this car was sold to Southern Pacific and was scrapped in 1930. *Solomon collection*

The Cumberland & Pennsylvania Brill gas-electric F101 had space for a Railway Express Agency as well as a passenger compartment toward the rear of the car. Brill was one of the largest suppliers of gas-electrics and offered a variety of standard models. *Solomon collection*

8572. Motor Car and Depot, Greeley, Colo.

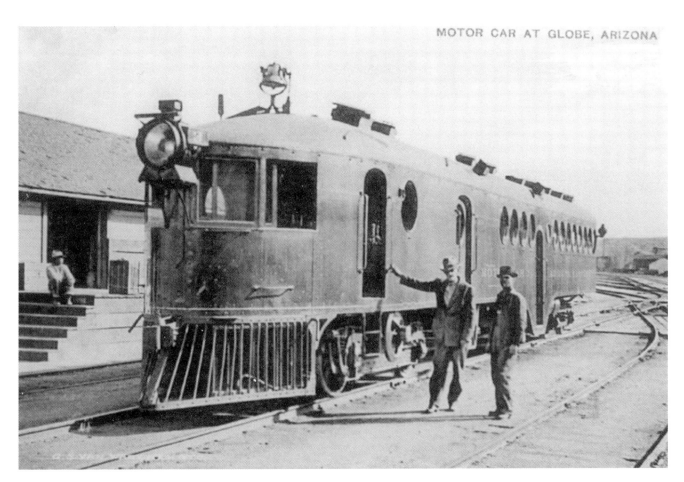

and mechanical transmissions proved McKeen's undoing. The McKeen cars tended to suffer from poor reliability and were remembered for their rough starts and burned-out clutches. Ultimately, UP assumed the company's assets. Although commercially unsuccessful, McKeen's futuristic "wind-splitters" are viewed as a precursor to internal combustion engine streamlined trains of the 1930s.

While McKeen was trying to overcome problems with his pseudo-aerodynamic cars with their cantankerous mechanical transmissions, a variety of other companies advanced gas-electric designs, colloquially known as "Doodlebugs." Among the most influential commercial producers was General Electric, which developed its self-propelled gas-electrics in tandem with its heavy railroad electrification business and supplies for street railway and interurban electric cars. From 1906 until 1914, GE sold eighty-eight gas-electric cars. Although GE wasn't the most prolific gas-electric supplier, its technological developments were later key for diesel-electric locomotive advancements.

Picking up where GE and others left the railcar was the Electro-Motive Engineering Corporation formed by Harold L. Hamilton and Paul Turner in 1920. These visionaries aimed to develop a prosperous railcar business by overcoming technical inadequacies that had limited the reliability and potential of earlier designs.

By the mid-1920s, Electro-Motive had become one of the most successful gas-electric railcar designers, selling dozens of railcars to railroads across North America. Significantly, Electro-Motive was strictly an engineering firm that owned no factories and so contracted all manufacturing to established railroad suppliers and manufacturers. In 1930, automotive giant General Motors bought Electro-Motive and, over the next couple of decades, transformed the company into the leading builder of diesel-electric locomotives.

This former Lancaster, Oxford & Southern wooden-bodied rail motor car was built by the Sanders Machine Shop in 1915. Now a diesel-powered car, it survives on Pennsylvania's Strasburg Rail Road. *Brian Solomon*

↖ Hoosac Tunnel & Wilmington was a short line that connected with Boston & Maine near the east portal of the famous 4.75-mile (7.6 km) Hoosac Tunnel in northwestern Massachusetts. In 1923, it bought a 9-ton railbus and trailer from the Four Wheel Drive Auto Company of Clintonville, Wisconsin. *Solomon collection*

↙ Kansas City Southern affiliate Louisiana & Arkansas operated this classic gas-electric with an unpowered trailer. The powered unit featured a crew cab, compartments for baggage, a Railway Express Agency (which handled packages and small shipments), and a U.S. Mail operated by Railway Post Office, while passengers were accommodated in the trailer. *Solomon collection*

↑ Two examples of Pennsylvania Railroad's self-propelled railcars meet at Princeton Junction, New Jersey, at 7 a.m. on May 29, 1959. On the left is a Pullman-built car, originally powered with a gasoline engine and later repowered with a diesel. On the right is one of the railroad's standard owl-eye MP54 electric multiple units working the branch to Princeton. *Richard Jay Solomon*

Pennsylvania Railroad K4s Pacific 4-6-2

The 4-6-2 Pacific emerged in the early twentieth century as a powerful, fast locomotive ideal for express passenger service. The most famous and one of the finest examples of the type was Pennsylvania Railroad's handsome big-boiler K4s Pacific. Confusingly, PRR included that small *s* as part of its locomotive designation, not to infer plural but to distinguish it as a locomotive with superheating at a time when many steam locomotives were not so equipped.

The 4-6-2 wheel arrangement had been variously applied by locomotive builders since the 1880s without notable success. The type came into the limelight in 1902, when Alco combined a wide firebox supported by a radial trailing truck in an order for 4-6-2 passenger locomotives for Missouri Pacific; the type was named "Pacific" in recognition of this pioneer application. The name echoed the 4-4-2 Atlantic type introduced a few years earlier while embracing the spirit of America's westward focus.

The Pacific was developed to accommodate America's railroad traffic growing by leaps and bounds. Railroads were running longer, heavier, and faster trains that demanded greater power than the older types could attain. The Pacific's large boiler capacity, six-coupled drivers, and four-wheel leading truck delivered a nearly perfect equipment combination. It was the logical expansion of both the 4-6-0 and the 4-4-2 types widely employed in passenger service and presented a superior arrangement over the 2-6-2 Prairie type, which suffered from stability problems at speed.

PRR was unusual in its scientific approached to locomotive design; it carefully tested new concepts before introducing them on a wide scale. Pennsylvania Railroad experimented with the Pacific type beginning in 1907, while continuing to refine its 4-4-2 Atlantic design as a fast, powerful passenger engine. It soon recognized that it needed more fast, powerful engines to accommodate its ever-growing swell of passenger traffic plus

↙ A classic locomotive in a classic pose: Pennsylvania Railroad K4s Pacific 5377 was photographed "rods down" at East St. Louis, Illinois, on October 28, 1944. Traditional steam locomotive photographers preferred to show locomotives with the rods in the lowest position to best portray mechanical equipment. *R. A. Frederick, Solomon collection*

↗ Among the final assignments for PRR's K4s Pacific locomotives was on the New York & Long Branch route between the end of the electrified zone at South Amboy and Bay Head Junction, New Jersey. PRR K4s 612 races along with a NY&LB train in the mid-1950s. *Solomon collection*

→ On December 7, 1945, PRR's *Jeffersonian* had eleven passenger cars and required double-headed K4s Pacifics 5493 and 5421. While a pair of K4s made for an impressive sight, operating two locomotives required two crews, which greatly increased costs. In the 1940s, PRR developed its streamlined 4-4-4-4 Class T1 to eliminate passenger double heading. *Unknown photographer, Solomon collection*

PRR K4s 1188 was photographed at Cincinnati, Ohio, on August 10, 1948. This engine featured an unusual valance atop the boiler, presumably as a smoke deflector to keep the cab free from swirling soot and steam. *J. R. Quinn, Solomon collection*

requirements hauling heavier steel equipment mandated by its New York Penn Station and trans-Hudson tunnels then under construction. PRR's Juniata Shops at Altoona built a large number of Pacifics between 1910 and 1913.

Meanwhile, on its own, locomotive manufacturer Alco pushed the limits of Pacific design, introducing in 1911 the experimental "super Pacific" No. 50,000, incorporating key modern innovations. This powerful engine inspired PRR designers, who in 1914 blended the most effective qualities of its E6s Atlantic and its own experimental 4-6-2 designs with Alco's innovations to create its own super Pacific, Class K4s. This has been deemed the finest North American example of the type, featuring 80-inch (203.2 cm) drivers with the large barrel Belpaire boiler and boxy firebox, which characterized most of PRR's twentieth-century steam locomotives. The K4s produced ample steam, delivering substantially more power than earlier Pacific locomotives. After a period

of testing and refinement during World War I, PRR began mass producing the engine at its Juniata shops, while also contracting Baldwin to build K4s to its specifications. Ultimately, PRR operated 425 examples of the type.

PRR was America's largest passenger carrier. While poster images of its K4s Pacific whisking the railroad's flagship New York–Chicago *Broadway Limited* along the Juniata River captured the public imagination, in practice, PRR's K4s fleet hauled a great variety of passenger trains across the system. And when diesels came after World War II, the K4s survived working less glamorous assignments. In 1957, they finished their days working suburban trains on New Jersey's New York & Long Branch to Bayhead Junction. Two of the class were preserved. One is displayed at the Railroad Museum of Pennsylvania at Strasburg; the other has been involved in a prolonged restoration effort.

↑ The K4s, with its large boiler and classic heavy lines, made it one of America's most recognizable steam locomotives. *Unknown photographer, Solomon collection*

→ PRR 3678 was one of four K4s dressed in this serviceable streamlined shrouding in 1940. It was seen leading the St. Louis–New York *Jeffersonian* along with a non-streamlined sister engine. *Unknown photographer, Solomon collection*

Gresley 4-6-2 Pacifics

Compared with North American railways, British railways were slower to embrace the 4-6-2 Pacific type as exemplified by the Pennsylvania Railroad K4s (see page 52), and instead continued to refine the 4-6-0 as an express passenger engine. However, the Pacific type was destined to make its mark in Britain in the 1920s.

The Big Four grouping of 1923—Britain's strategic national railway consolidation that merged most smaller railways into four regional networks—created the London & North Eastern Railway (LNER) and melded the Great Northern Railway with the North Eastern Railway, which put the East Coast route (the direct fast line between London's King Cross Station and Edinburgh, Scotland)

under common ownership. Immediately prior to this grouping, both of these key LNER components adopted the Pacific type as a means for better handling longer East Coast express passenger trains. Nigel Gresley had come to LNER from the Great Northern, where he had recently introduced a novel three-cylinder Pacific design that he classed A1. As LNER's chief mechanical officer, he continued to refine and perfect his Class A1 4-6-2 type.

In Gresley's three-cylinder engine, all the cylinders powered the second driving axle to avoid the difficulties associated with divided drive (cylinders powering different axles). The outside cylinders followed the well-established practice of using main drive rods to power the second driving axle with Walschaerts valve

↙ London & North Eastern Railway's Pacific No. 4472 *Flying Scotsman* leads a recreated Flying Scotsman at York on July 4, 1999. LNER's famous train and its even more famous locomotive are often confused with one another. *Brian Solomon*

→ A detailed view of locomotive 4472 *Flying Scotsman* shows the plaque commemorating the locomotive's record-breaking long-distance nonstop run in New South Wales, Australia, in 1989. *Brian Solomon*

gear to govern steam admission to the cylinder. However, the central cylinder faced space constraints and so powered the second set of driving wheels via a crank axle that required this cylinder to be inclined. Normally, valves were located directly above the cylinder they regulated, but to keep the central cylinder valve on a level plane with the outside cylinder, it was necessary to place the central valve adjacent to the cylinder rather than above it; this arrangement precluded the application of a full Walschaerts system for the central cylinder. So Gresley adapted a conjugated gear that bears his name, which derived motion by using levers to combine the events of the two outside Walschaerts valve gear systems to regulate the central valve events. Gresley's A1 Pacific proved to be an excellent express engine, highly regarded by LNER.

A locomotive is a piece of machinery, while a named train is a scheduled service.

When LNER named A1 1472 (later renumbered 4472) *Flying Scotsman*, it inadvertently flummoxed generations of railway observers, who often confused the steam engine with LNER's famous scheduled service called the Flying Scotsman that ran between London and Edinburgh.

Engine 4472, *Flying Scotsman*, became famous through its connection with its namesake train and from its widely publicized test run in 1934, where it topped 100 mph (160.9 km/h). In later years, the engine earned even greater fame as a popular mainline excursion locomotive, which toured railroads in the United States and Australia as well as British main lines.

Gresley continued to refine the Pacific type, making it more powerful and faster. In 1928, he built the first of his A3 Pacifics; LNER ultimately built twenty-seven new A3s while rebuilding older A1s to A3 specifications. Then

In 1969, locomotive 4472 *Flying Scotsman* made a widely publicized tour of the United States, when the engine was modified with the addition of a bell and "cow catcher" pilot. It was seen here under wire on Penn Central's former Pennsylvania Railroad multiple track main line. *Photographer unknown, Solomon collection*

in the mid-1930s, with the advent of high-speed German and American diesel streamliners, LNER was encouraged to introduce fast streamlined trains on the East Coast route. This resulted in the streamlined A4 Pacific in 1935, with wind-resistant shrouds covering the highly refined Gresley 4-6-2 Pacific type.

LNER introduced its fully streamlined, lightweight express train called *Silver Jubilee* on September 30, 1935. Named to commemorate the twenty-fifth year of Britain's King George V, this ran between New Castle and London on a four-hour sprint. Over the next three years, LNER built thirty-five streamlined A4 Pacifics. In 1938, the recently

knighted Gresley was delighted to have the one-hundredth Pacific of his design named in his honor.

Most famous of all the A4s was *Mallard*, which gained world fame in a special speed run on July 3, 1938, when it hit 126 mph (202.8 km/h)—claimed as the fastest recorded steam run in the world. Gresley's Pacifics worked mainline trains until the 1960s, when most were withdrawn and retired. Several have been preserved and occasionally operated, including the famous *Flying Scotsman* and the *Sir Nigel Gresley*. Famous speedster *Mallard* is a central display at Britain's National Railway Museum in York.

↑ London & North Eastern Railway's *Mallard* holds the world record for fastest documented steam run. Today, this engine is proudly displayed at the National Railway Museum at York, England. *Brian Solomon*

↓↓ Among the preserved LNER Class A4 streamlined Pacifics is *Union of South Africa*, seen here restored to its British Railways appearance when it carried the number 60009. *Brian Solomon*

Berkshire 2-8-4

n the early 1920s, the Lima Locomotive Works was America's newest and smallest of the three large commercial locomotive builders. To boost its market share, Lima's engineers set out to design a better locomotive to give it an advantage over locomotives built by established builders, Baldwin and Alco.

At that time, American railroads were saturated with freight traffic yet were facing increased competition from trucks skimming away the most lucrative business. To counter this rubber-tire invasion, railroads needed to run faster trains with lower operating costs. Lima's Will Woodard was among the most talented locomotive designers in the United States; he previously worked for both Alco and Baldwin before leaving the big builders to work for upstart Lima in 1916. There, he put a

visionary plan into motion, which ultimately affected the entire locomotive industry. Instead of merely designing bigger, heavier locomotives, he crafted more powerful engines that demonstrated markedly better performance without a dramatic weight increase. Significantly, by keeping weight and proportions within established limits, these locomotives enabled railroads to haul more tonnage faster, without needing to invest in expensive line upgrading.

Among the lines leading the change for better motive power was New York Central with its four-track "Water Level Route," one of the busiest freight corridors in the United States. NYC was one of the pioneers of fast freight; its concept was to move trains of traditional lengths faster, rather than just moving bigger trains. One difficult area for New York

Boston & Albany A1a class Berkshire No. 1400 works east with tonnage near Natick, Massachusetts, in 1944. *Robert A. Buck collection*

Central was its heavily graded Boston & Albany route. It was here that Lima demonstrated its superpower concept with a 2-8-4 prototype. The first big locomotive to adopt this wheel arrangement, it featured a four-wheel load-bearing trailing truck that allowed for construction of a substantially larger firebox in order to heat ample quantities of steam. The locomotive was an advancement on Lima's super modern H10 2-8-2 Mikado built for New York Central a few years earlier. Like the H10, the 2-8-4 benefited from recent advancements in steel production that created new high-tensile alloy steels at reasonable cost. Heat-treated chrome vanadium steel was used for side and main rods, which reduced reciprocating weight and contributed to smoother running.

The combination of a bigger firebox with corresponding improvements in the boiler and better running gear resulted in substantially better performance: the 2-8-4 could haul great tonnage, faster and with less fuel than even the best 2-8-2s of the time. The 2-8-4 wheel arrangement was named the Berkshire type, and New York Central placed three orders for service on the B&A. Other railroads followed suit. Later 2-8-4s featured taller drivers for faster running. Among the finest were big Limas built for the Nickel Plate Road a generation after B&A's. Lima expanded the 2-8-4 into the 2-10-4, first sold to the Texas & Pacific and known as the Texas type. By the late 1920s, the superpower concept had taken hold and all three builders had emulated Lima's new pattern, leading to a succession of bigger and more powerful steam locomotives.

Lima wasn't the only manufacturer that built 2-8-4s: Nickel Plate Road No. 802 was one of thirty-two Alco 2-8-4s built for Wheeling & Lake Erie, seen here in the mid-1950s working a heavy freight in central Ohio. *J. William Vigrass*

← A pair of Boston & Albany Berkshires at work was a mighty sight. After taking water at West Brookfield, Massachusetts, two 1400 Berkshires resume their march east. *H. W. Pontin, Robert A. Buck collection*

↙ The most famous Nickel Plate Road Berkshire is engine 765, owned by the Fort Wayne Railroad Historical Society in Indiana. This classic highstepping Lima 284 has made a career as an excursion locomotive. On July 30, 1988, it roared west along the old Erie Railroad at Tuxedo, New York. *George W. Kowanski*

↓ A pair of Western Maryland F7s lead Nickel Plate Berkshire 759 upgrade toward Frostburg, Maryland, at the famous Helmstetters Curve on October 17, 1970. *George W. Kowanski*

New York Central Hudson 4-6-4

New York Central's Water Level Route was one of the United States' principal arteries of commerce in the nineteenth and twentieth centuries. This largely four-track ran from New York City to Chicago following the Hudson River north toward Albany, then west along the route of the old Erie Canal through a heavily industrialized region via Schenectady, Syracuse, Rochester, and Buffalo, New York, then along the shores of Lake Erie through Erie, Pennsylvania, Cleveland and Toledo, Ohio, and across northern Indiana.

The New York Central System was the second-largest American railroad and vied for transport supremacy with its archrival—the Pennsylvania Railroad. The two giants competed for territory, traffic, and prestige. In the railroads' golden age, a line's flagship passenger train was a formidable advertising tool. Central's best-known and most impressive train was the *20th Century Limited*, which in its heyday raced nightly over the length of the Water Level Route in just over sixteen hours, from Grand Central in New York City to LaSalle St. Station in Chicago, as an exclusive extra-fare, first-class Pullman sleeper consist. It was the pride of the railroad's "Great Steel Fleet" that included a host of named limited trains connecting cities across the Northeast and Midwest.

As the penultimate American passenger carrier, Central took great pride in its express engines and pushed development of ever larger and more powerful machines. By the mid-1920s, its burgeoning passenger traffic saw the

At Montrose, New York, on April 7, 1947, New York Central J-1b 5244 leads the first section of train 44, the *New York Special*. Edward L. May, New York Central System Historical Society

4-6-2 Pacific at its practical limits of size and power, and because Central was an old route built to early standards, with an unusually restrictive loading gauge that constrained the maximum width of locomotive boilers as compared with more generously built lines, it needed a bold solution. Paul Kiefer, the railroad's chief locomotive designer, brought the design of six-coupled steam to the next level by introducing the 4-6-4 type. This featured a four-wheel trailing truck (like that first used on the 2-8-4 Berkshire type in 1924) to facilitate construction of a substantially larger firebox and a high-capacity tender to supply sustained quantities of steam to maintain top track speeds with Central's heavy consists.

Working with Alco, the first Hudson, Class J-1, made its public debut on Valentine's Day in 1927. This was an exceptional locomotive followed by another 204 engines from Alco. The J-1s had an unadorned utilitarian appearance but nevertheless were handsome and well-balanced machines that exuded power. They allowed the railroad to lengthen trains while meeting tight schedules and reducing operating expenses.

Building from this success, in 1935, Kiefer refined the Hudson specification to produce the Class J-3a "Super Hudson." He fine-tuned cylinder sizes, increased boiler pressure, and employed modern lightweight alloy steel for reciprocating parts along with roller bearings and modern drive wheel designs. This gave the J-3a superior performance, fuel efficiency, and reliability. The most famous of the type were sleek streamliners featuring classy art deco shrouding designed by Henry Dreyfuss for use on the new 1938 *20th Century Limited*. Yet their glory days were short, and the J-3as succumbed to diesel power after World War II. In the 1950s, Central's management lacked any nostalgia for its technological achievements, and when the diesels took over, they scrapped every last Hudson.

New York Central's J-1 class Hudson was a formidable machine with lots of boiler power. Central's J-1b 5229 was leading an excursion and was polished for the occasion. *Solomon collection-*

↑↑ New York Central's Great Steel Fleet included numerous named trains, such as No. 67, *Commodore Vanderbilt*, named to honor the railroad's founder. It was pictured led by J-1b 5215 near the Bear Mountain Bridge at Manitou, New York, on June 16, 1940. *Edward L. May, New York Central System Historical Society*

↑ New York Central's finest Hudsons were the J-3as built in 1937–1938. Engine 5441 scoops water on the fly at Tivoli, New York, on August 30, 1941. *Edward L. May, New York Central System Historical Society*

It was an era of art deco elegance. Streamlined J-3a Hudson 5447 races north along its namesake river at Manitou, New York, with the 1938 *20th Century Limited*. Edward L. May, New York Central System Historical Society

Northern 4-8-4

The rapid progression of American steam power saw passenger types progress from 4-4-2 to 4-6-2 to 4-8-2, and a corresponding progression of freight designs from 2-8-0 to 2-8-2 to both 4-8-2 and 2-8-4s in roughly the same time frame. This pointed the way for a common wheel arrangement that was equally well suited to both freight and passenger traffic. Yet the first 4-8-4 was a curious aberration, developed for Northern Pacific to burn low-grade Montana coal.

In 1926, Northern Pacific looked to Alco for a big passenger locomotive that could make better use of cheap Montana "Rosebud" coal that was in ample supply along its lines. This resulted in the first 4-8-4 design, which Alco adapted from the 4-8-2 Mountain type to feature a larger firebox with a broad grate in order to burn low-yield Rosebud coal.

The type became known as "Northern" in honor of NP, and soon three other North American railroads adopted the type. Ironically, the Northern name was never universally accepted. In fact, no other steam locomotive type enjoyed more names than the 4-8-4. On the heels of NP's pioneer, Alco built 4-8-4s with high drivers for Lackawanna, known as "Poconos," and Canadian National focused on lightweight 4-8-4s for freight and passenger service across its system that it called "Confederations." Santa Fe was the first to buy a Baldwin-built 4-8-4.

The most advanced and most famous 4-8-4s are among America's best-known steam locomotives. The list includes semi-streamlined

↓ Northern Pacific was the first to adopt the 4-8-4 wheel arrangement, which was essentially an adaptation of the 4-8-2 Mountain type, configured with an usually large firebox to burn low-quality lignite. *Solomon collection*

Santa Fe was a large buyer of the 4-8-2 Mountain type, and in 1927 it was the first to buy a Baldwin-built 4-8-4. It embraced this type for general road service, continuing to order these through World War II. The first 4-8-4 3751 has been preserved and restored. It is seen here on former home rails passing San Clemente, California, on the shore of the Pacific. *Brian Solomon*

Lima-built engines for Southern Pacific—best remembered by its No. 4449, which was preserved and restored to mainline excursion service in the 1970s; Union Pacific's highdriver 800 class iconified by engine No. 844, the last of its kind built and which was never retired at the end of steam, continuing to perform on excursions to the present day; and Norfolk & Western's elegantly streamlined J class engines as typified by preserved engine No. 611.

← → In 1945, Reading Company built a fleet of Class T-1 4-8-4s by recycling boilers and fireboxes from some older 2-8-0 Consolidations. In 1959, after it dieselized revenue freight, it began operating popular Reading Iron Horse Rambles excursions with surviving T-1s. *Richard Jay Solomon*

↘ Canadian National acquired the greatest number of 4-8-4s, some 203 locomotives, including those of its American affiliate, Grand Trunk Western. CN's 4-8-4s were handsome machines and designed to work both freight and passenger trains. These had relatively light axle weights compared with the 4-8-4s used on many lines in the United States, and this gave the Canadian locomotives greater route availability because they could be assigned lines with lightly built bridges. CN 4-8-4 6167 works an excursion in June 1961. *Richard Jay Solomon*

Flying Hamburger

Germany was an early innovator of high-speed train development and set several important precedents that shaped locomotive and train development in Europe and America. By 1903, Germany was pushing top speeds to new limits. The Berlin-Zossen speed demonstrations showed that electric trains could reach more than 130 mph (209.2 km/h). In the 1930s, it pushed the limits further with its bizarre looking "Rail Zeppelin," a single lightweight car that easily reached 100 mph (160.9 km/h) and was reported to have reached 143 mph (230.1 km/h) in a high-speed test.

More significant was the development in 1932 by Wagen und Maschinenbau AG of its diesel-electric, aerodynamically designed, articulated two-piece railcar. Where early diesel engines were slow, ponderous, and heavy, this train was powered by a pair of modern lightweight 410-horsepower Maybach diesels. The futuristic lightweight aerodynamic train design benefited from advanced automotive and aircraft innovations, while the shape of the train stemmed from exhaustive wind tunnel experiments at the Zeppelin Works on the shore of Lake Constance in Friedrichshafen that aimed to reduce wind drag and to allow the train to travel faster with lower energy consumption. The effects of wind resistance increase as a body moves faster, and with intended speeds in excess of 100 mph (160.9 km/h), aerodynamics was deemed to be of great importance.

On May 15, 1933, the train entered revenue service on Deutsche Reichsbahn (German State Railways) between Berlin and Hamburg as the *Fliegende Hamburger* (*Flying*

↙ In 1901, the Zossen–Marienfelde military railway near Berlin was converted into a high-speed advanced railway testing site, specially wired with a high-voltage, three-phase alternating current system, built by Siemens and Halske, and AEG. Tests in 1903 demonstrated that trains could hit speeds up to 130 mph (209.2 km/h). *Solomon collection*

↗ A restored *Flying Hamburger* railcar on display at Leipzig Hbf (main station). Germany's sleek, aerodynamically designed diesel-electric railcars inspired development and set the pattern for similar diesel-articulated trains in the United States in the mid-1930s, notably Union Pacific's *Streamliner* and Burlington's Budd-built *Zephyr*. *Brian Solomon*

Elektrischer Triebwagen
auf der Versuchsstrecke Marienfelde—Zossen, ausgerüstet von Siemens & Halske 1901—1903
(Erreichte am 23. Oktober 1903 eine Geschwindigkeit von 206,7 km/h)

Hamburger). It was the fastest regularly scheduled train in the world and zipped over its 178-mile (286.5 km) run in just 2 hours 18 minutes, making an average speed of 77.4 mph (124.6 km/h). It often reached more than 100 mph (160.9 km/h) for extended periods to maintain its ambitious running time and was much faster than traditional steam-hauled trains. While diesel engines and lightweight aerodynamic design enabled the train to go fast, equally important was its advanced braking, including automatic train stop, which allowed it to stop safely within the limits of existing infrastructure.

The success of this train encouraged Deutsche Reichsbahn to create a whole network of fast daily trains connecting German cities, and it ordered seventeen similar diesel trainsets, of which thirteen were two-piece railcars and four were three-piece railcars. By 1935, Deutsche Reichsbahn had twelve of the world's fastest scheduled trains. While the fast services were discontinued with the advent of World War II, some of the railcars survived the war and continued to work into the 1950s. A few were preserved, with the original *Fliegende Hamburger* having been displayed at the Nurnberg railway museum.

↙ The historic Deutsche Reichsbahn insignia on the side of a preserved *Flying Hamburger* railcar. *Brian Solomon*

← In 1932, Wagen und Maschinenbau AG built the world's first high-speed diesel train using an articulated, two-piece, streamlined diesel-electric railcar powered by a pair of twelve-cylinder, 410-horsepower Maybach engines. Its aerodynamic design and modern automotive and aircraft manufacturing techniques produced a radical new type of train. *Brian Solomon*

→ In 1935, special-issue German stamps commemorated one hundred years of German railways. At the bottom, modern streamlined trains included *Flying Hamburger* diesel railcars at left, and a streamlined 4-6-2 Pacific (right) contrasting with *Der Adler* (*The Eagle*), the first locomotive to operate on German rails (top left). *Hundert Jahre Deutsche Eisenbahnen, Solomon collection*

↓ One of Germany's aerodynamic two-piece articulated railcars glides along at speed. These fast, lightweight, diesel powered trains inspired development of similar trains in the United States including the famous Burlington Zephyr of 1934. *Hundert Jahre Deutsche Eisenbahnen, Solomon collection*

London Midland & Scottish Black Five 4-6-0

Perfect is a word rarely applied to locomotives, but one of the most perfect British locomotives was the so-called "Black Five," a standard 4-6-0 type designed by the London Midland & Scottish Railway (LMS) in 1934. This versatile, reliable, handsome, compact, and powerful locomotive earned high praise from both operating men and maintenance staff. This engine was among LMS standardized locomotives that were originated by the railway's chief mechanical engineer, William A. Stanier, who introduced a policy of integrated standard designs. It was one of the most numerous British steam locomotive types; following nationalization, it was widely used on lines across the United Kingdom.

In the 1930s, Stanier introduced a policy at the LMS for an integrated fleet of standardized locomotives. Not everyone at the railroad was enthusiastic about his policy, and he had to overcome internal resistance before introducing five standard types, including his two cylinder 4-6-0 built for both freight and passenger work, a type that had no precedent on the line. This type earned the colloquial moniker "Black Five" because of the LMS 5P5F classification and its basic black paint.

Stanier had come to the LMS after a long career with Britain's Great Western Railway (GWR), which was famous for its fleet of 4-6-0s. He patterned the Black Five after GWR's successful King Class 4-6-0. There were several differences between the two types: the

Dressed in BR black paint, former LMS Stanier–designed Black Five approaches Highley on the preserved Severn Valley Railway. *Brian Solomon*

Black Five 45110 approaches the station at Arley on Britain's scenic Severn Valley, one of dozens of thriving heritage railways in the United Kingdom. *Brian Solomon*

Kings used Stephenson inside valve gear, while the Black Five used Walschaerts outside gear.

Engine 5000 was constructed in 1934 at LMS's Crewe Works, and over the next seventeen years, LMS took delivery of 842 Black Fives; many were built at the railway's shops in Crewe, Derby, and Horwich, and because it was a standard design, large numbers were also built commercially. The success of the type saw many of them survive until the end of British steam operations in 1968. Luckily, nineteen of the 5P5F class escaped scrapping and operate today on British heritage railways.

↖ Black Five 45110 now resides on the Severn Valley Railway. This is one of 842 similar locomotives and is often claimed as one of the final British Railways' steam locomotives in revenue service. *Brian Solomon*

← The clean lines and well-balanced design of Stanier's Black Five, combined with the exceptional numbers of the type built, have contributed to the type's longevity and continued popularity among the legions of preserved steam locomotives in the United Kingdom. *Brian Solomon*

↑ Similar to LMS's Black Five were British Railways' Standard 5 4-6-0s, which were patterned after the older type. Locomotive 73082 *Camelot* resides on the Bluebell Railway. Located in the former Southern Region southeast of London, Bluebell was the first standard gauge heritage railway in the United Kingdom. *Brian Solomon*

Pennsylvania Railroad GG1

Most famous of America's steam-era electric types was Pennsylvania Railroad's streamlined GG1. The majesty of these elegant machines cannot be fully conveyed in photos. These powerful and exceptionally capable electrics served PRR and its successors for the better part of five decades. The stepped whine of a G's traction motors as it accelerated, the low guttural blast of its air horn, and the clattering of its multiple axles were born of the steam era yet outlived steam by a generation.

During the 1920s and 1930s, PRR embarked on America's most intensive mainline high-voltage AC electrification by wiring its busy New York–Philadelphia–Washington, D.C., routes. Ultimately, this extensive project included routes to Harrisburg, Pennsylvania, and various branches and secondary freight cutoff routes. Initially, PRR built fleets of double-ended boxcab electrics that used adaptations of common steam locomotive wheel arrangements. Most numerous was the P5, with three sets of drivers patterned on the

Pacific type. However, PRR soon found serious flaws with its boxcab designs, especially when operated at speed. The P5s suffered from excessive lateral sway and developed axle cracks. Furthermore, a serious grade-crossing accident involving a P5 demonstrated the dangers of frontal cab placement.

In early 1933, PRR established a test track at Claymont, Delaware, to develop a superior class of electrics. It borrowed a New Haven Railroad Class EP3 boxcab that featured an articulated wheelbase with a 2-C+C-2 arrangement. This provided a better ride quality while enabling power to be distributed over more axles. Then, in 1934, PRR built two prototypes: a Class R1 that featured a rigid base with 2-D-2 wheel arrangement, and a Class GG1 patterned on New Haven's EP3. Both locomotives used a revised carbody arrangement, which featured central cabs protected by long tapered nose sections using riveted streamlined sheet-metal skin credited to the design Westinghouse's Donald Dohner. The GG1 arrangement prevailed, and PRR selected

On Pearl Harbor Day in 1958, then-ordinary PRR GG1 4935 races *The Admiral* eastbound through Overbrook, Pennsylvania. Two decades hence, Amtrak selected this locomotive for special treatment, and it was overhauled and restored to its classic appearance. Today, 4935 is proudly displayed at the Railroad Museum of Pennsylvania in Strasburg.
Richard Jay Solomon

this as its standard electric. Before the GG1 entered production, PRR hired French-born pioneer industrial designer Raymond Loewy to improve the locomotive's appearance. He suggested a variety of subtle physical changes, including using welded body construction along with various exterior refinements and the now-iconic Brunswick green with gold pinstriped "cat's whiskers" paint livery.

Ultimately, PRR ordered 138 production GG1s that were variously constructed between 1935 and 1943 by GE at Erie, Pennsylvania; Baldwin at Eddystone, Pennsylvania; and its own Juniata Shops in Altoona. These were rated at 4,620 horsepower continuously, though they could deliver short bursts of higher horsepower and were capable of whisking an eighteen- to twenty-car passenger train along at 90 mph (144.8 km/h). In the 1970s, Amtrak and Conrail inherited many GG1s and operated them until 1981. The final thirteen worked for NJ Transit between New York Penn Station and South Amboy until October 1983. Several examples have been preserved and cosmetically restored.

Pennsylvania Railroad GG1 electric 4868 is turned on the loop track at Sunnyside Yard in Queens, New York, in March 1961. For almost fifty years, GG1s were serviced at Sunnyside Yard. *Richard Jay Solomon*

In the 1950s, PRR
treated a few of its GG1s
to a variation of the
Loewy paint scheme,
substituting Tuscan red
for Brunswick green
to match its passenger
fleet. On June 8, 1958,
4910 rolls through
Trenton, New Jersey.
Richard Jay Solomon

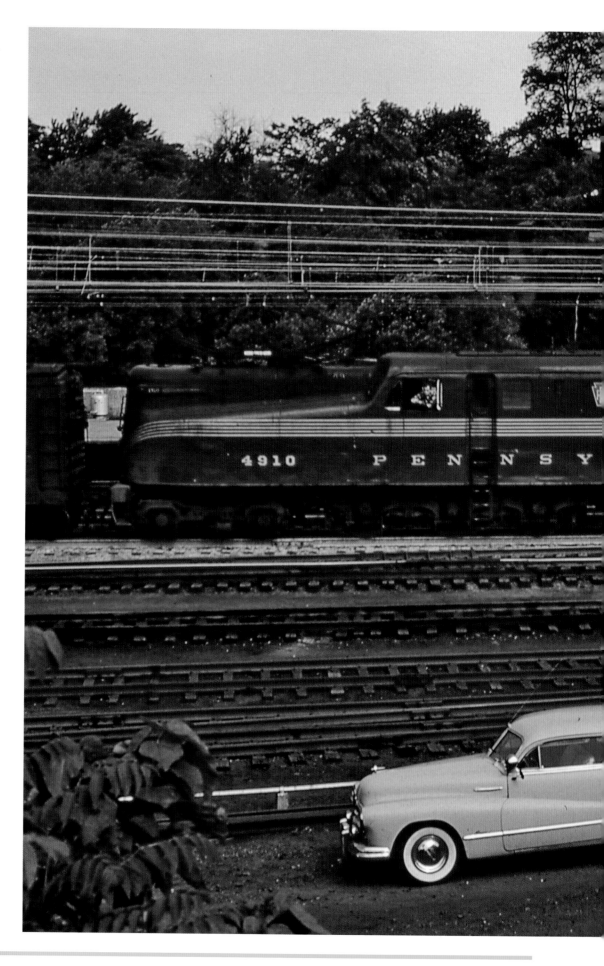

Type 12 Atlantic 4-4-2

Among the most unusual looking and fastest steam locomotives in Continental Europe was the Belgian Type 12 of 1939. Belgium has the oldest and densest mainline railway network on the continent. Since the country has two primary languages owing to its dichotomous culture, the Belgian national railway is known both in Flemish as the Nationale Maatschappij der Belgische Spoorwegen (NMBS) and in French as Société National des Chemins de fer Belges (SNCB).

In the late 1930s, NMBS/SNCB aimed to add prestige and to speed up its premier express trains connecting the North Sea port at Ostend with Brussels—Belgium's capital and largest city. Ostend offered direct ferry connections with England, so the Ostend–Brussels route served as a primary corridor for travel between the United Kingdom and cities across Central Europe. Inspired by fast streamlined steam locomotives in North America, locomotive designer Raoul Wotesse emulated the success of Alco A1 4-4-2 streamlined Atlantics built for Milwaukee Road's superfast Hiawatha that connected Chicago–Milwaukee–Minneapolis/St. Paul, and the derivative 4-4-4 Jubilee type built for Canadian Pacific by Alco's Canadian affiliate Montreal Locomotive Works. He worked with French industrial designer André Huet—also known for streamlining French locomotives—to wrap the new Belgian locomotive with futuristic metal shrouds. Significantly, while these were aimed to reduce wind resistance at speed, open areas in the shrouds were retained to facilitate maintenance. This overcame a principal flaw

with earlier streamlined steam designs where all-over shrouding had the side effect of complicating basic maintenance and increasing operating costs.

Six Type 12 Atlantics, originally numbered 1201–1205, were built by Belgian manufacturer Cockerill at Leige in 1939. These featured exceptionally tall driving wheels to enable sustained fast running. Among the unusual elements of the design were inside cylinders powering the first set of driving wheels using inside rods and a cranked axle, yet featuring outside connecting rods between drivers. The first four locomotives used Belgian-designed Walschaerts valve gear (that had become one of the most common types of outside valve gear used by locomotives around the world), while the last two engines used two different systems of state-of-the-art poppet valves for precision control of steam admission to the cylinders.

In May 1939, engine 1202 made a well-publicized high-speed sprint from Brussels Midi to Ostend Quay, where it hit 103.5 mph (165 km/h) and maintained an average speed of 75.2 mph (121 km/h), establishing a European "blue ribbon" speed record. The locomotives' heyday was exceptionally short-lived since by September 1939, Europe was enveloped in World War II. After the War, the surviving Type 12s continued to work in express service. In September 1962, locomotive 1204, by then renumbered 12004, was withdrawn from traffic and preserved. Since 2015, this has been displayed at the Train World museum at Schaarbeek in Brussels, not far from where it was stabled during its operational years.

The surviving example of NMBS/SNCB's streamlined Type 12 Atlantic makes for a stunning display at Belgium's Train World museum in Brussels. The artificial smoke panel positioned above the smoke stack is designed for dramatic effect. *Brian Solomon*

Electro-Motive F-Unit

I n late 1939, following several years of intensive development, General Motors' Electro-Motive Corporation (soon to be renamed its Electro-Motive Division) introduced its groundbreaking model FT road diesel. The basic technology, encased in handsome streamlined locomotive contours, had been refined from EMC's earlier streamlined passenger service E-unit models. The FT clearly demonstrated to American railroads that a modern diesel-electric locomotive was sufficiently reliable, durable, capable, and flexible for a wide range of mainline work, including movement of the heaviest freight trains.

Significantly, General Motors had designed the FT to work as combinable "units" that contributed to the FT's great service flexibility. FTA units featured cabs, while the FTB units were cab-less "boosters," and in their original configuration were semi-permanently coupled as A-B pairs with some electrical components shared between the two units. The A-B sets produced 2,700 horsepower and were intended to rival the output of a 2-8-2 Mikado-type steam locomotive, which since World War I

had emerged as a common freight service workhorse. When combined as A-B-B-A sets, the FT produced 5,400 horsepower and equaled or exceeded the pulling power of even the most modern "superpower" steam. The FT demonstrated most of the advantages of a straight electric locomotive without the need for expensive electrification.

Following an extensive demonstration tour, the FT entered regular production, which occurred on the eve of American involvement in World War II. General Motors' timing was significant, as the war produced a profound effect on the application of the diesel—traffic soared, creating enormous demand for motive power while wartime restrictions on locomotive purchases mandated by the War Production Board greatly limited a railroad's ability to buy motive power, and especially diesels. So while the war limited sales, it gave GM's designers the opportunity to conduct extensive research and development of their locomotive's primary components. This allowed them to better understand the cause of component failures and refine and improve

When it was new, Electro-Motive's demonstrator FT model was a handsome streamlined machine boldly displaying "GM" on its nose and "Electro-Motive" on its sides. This radical new locomotive convinced many railroads that diesels were the future of American motive power. Once railroads sampled the FT, most never bought another steam locomotive. *Solomon collection*

their designs to make their locomotives more reliable.

While wartime service had demonstrated the functional superiority of GM's F-unit locomotive, after the war, GM implemented numerous design improvements to their postwar diesels, which rapidly propelled them as America's foremost locomotive manufacturer. The new F3 was offered with a variety of gearing options to allow railroads to tailor their diesels for intended service. While primarily a freight locomotive, the F3 could also be assigned passenger work.

Key to GM's diesel strategy was introducing engineering changes in batches rather than implementing piecemeal improvement while carefully engineering improvement to ensure component compatibility with older designs. Using this strategy, every few years GM's new locomotive models would supersede older ones in a calculated effort to improve reliability and output. In 1949, the F7 replaced the F3. Where both models were rated at 1,500 horsepower

per unit, the F7 had more durable components and overcame shortcomings associated with older designs. The FP7 was a model intended specifically for passenger service and featured greater steam heat capacity.

In 1954, GM introduced the 1,750-horsepower F9, which also featured a host of improvements to reliability and operation. However, by the mid-1950s, American railroads were moving away from streamlined carbody-style locomotives in favor of cheaper and more versatile road switchers better suited to bi-directional operation. By the late 1950s, while a few railroads continued to order F-units, most preferred road switchers, such as the popular GP9, which used most of the same basic components as the F but was arranged in a more versatile format. The last F-units were specialized dual-mode diesel-electric/third-rail electrics designed for the New Haven Railroad for operation into New York City's underground terminals. These employed the unusual B-A1A wheel arrangement.

New York Central was among buyers of Electro-Motive's FT diesels. FTA 1600 was photographed at Elkhart, Indiana, on March 18, 1945. *Robert A. Buck collection*

← Following its late 1939 debut, the pioneer Electro-Motive FT embarked on an eleven-month, 83,764-mile (134,805 km) tour of twenty Class I American railroads. This significant locomotive has been preserved and was photographed at Spencer, North Carolina, during a "Streamliners" locomotive gathering in 2014, where it was among the stars of the show. *Brian Solomon*

↙ In October 1964, battle-worn Boston & Maine F3 4228 leads joint Central Vermont Railway–B&M passenger train at White River Junction, Vermont. *Richard Jay Solomon*

→ Among the identifying features of the FT is its rows of four porthole windows on the sides of the carbody. *Brian Solomon*

↑ In 2010, New England's Pan Am Railways made a trade with Conway Scenic Railroad, swapping a GP38 and GP35 from its freight fleet for Conway's "Sisters," as the pair of former Canadian National FP9s were know. These were shopped as PAR1 and PAR2 for use on the company office car train, seen here exiting the east portal of Boston & Maine's Hoosac Tunnel in the midst of a February 2016 snow squall. *Brian Solomon*

↗ The final sixty EMD F-units were the most unusual. New Haven Railroad, which had ignored EMD during the early phases of its dieselization, turned to EMD in the mid-1950s for a fleet of dual-mode diesel-electric/ third-rail electrics for service into New York City's electrified terminals. This pair of FL9s works along New York Central third-rail territory in the Bronx, New York, in 1961. *Richard Jay Solomon*

↑ Although EMD F-units represented the face of North American dieselization, these classic locomotives were relatively rare on some lines, such as Canadian National's American affiliate, Grand Trunk Western. Where railroads such as New York Central bought Fs in the hundreds, GTW's twenty-two F3As built in 1948 were an anomaly in its fleet. *Richard Jay Solomon*

→ In May 2014, North Carolina Transportation Museum's Spencer Shops hosted a streamliner event that gathered more than two dozen EMD E- and F-unit diesels from around the country. Here, a former Southern Railway FP7 6133, Wabash F7A 1189, and a former Bangor & Aroostook F3A dressed as Lackawanna 663 posed for night photos. *Brian Solomon*

Vladimir Lenin Electrics

The first electrified railway in the Soviet Union (USSR) was a relatively short suburban passenger operation in Baku, the capital of Azerbaijan Soviet Socialist Republic. The USSR viewed electrification as the most effective means of increasing route capacity, so in the 1930s it began wiring strategic routes using 3,000-volt DC overhead catenary, including the graded Trans-Caucasus line over Suram Pass. This difficult route had previously employed 0-10-0s and an unusual fleet of Fairlie-type steam locomotives. The USSR's electric railway technology was in its infancy, so it imported eight General Electric–built electrics similar to some of the most advanced types of the period built for the New York Central and New Haven Railroads in the United States. The USSR's Type S electrics rode on C-C trucks with series-wound nose-suspended traction motors. About the same time, the USSR refined its own electric locomotive design designated VL19; the letters VL honored Vladimir Lenin, the USSR's early leader and ideological figurehead, while the number 19 reflected its 19-ton axle weight. The VL19 type fared poorly compared with the GEs, so further electric locomotive development focused on advancing GE design principles. This resulted in the VL23 of 1938 and continued to influence Soviet electric locomotive design for the next quarter century.

Although the USSR aggressively extended its 3,000-volt DC electrification after World War II, in the mid-1950s, it began to experiment with high-voltage AC electrification. It settled on 25kV at 50Hz as its new preferred mainline standard yet continued to expand DC electrification where this was already dominant. The USSR's first mass-produced AC electric was the N60 type built at the Novocherkassk Works beginning in 1957. Like the early DC-powered units, these six-motor electrics rode on C-C trucks. Early examples used mercury-arc rectifiers, but difficulties with this technology resulted in later locomotives employing the more reliable silicon diode rectifier system. In both scenarios, 25kV AC was converted to direct current to power series-wound nose-suspended traction motors. An estimated 2,600 N60s were built, making this type among the most numerous electric locomotives of the period.

In 1961, the USSR introduced the N80 type, a two-section heavy electric using a B-B+B-B wheel arrangement, specifically intended for heavy freight work. Like the N60, early examples of the N80 were mercury-arc rectifiers, but most production locomotives built after 1963 used the silicon diode system. These were exceptionally powerful machines that produced nearly 8,400 horsepower. More than 2,400 were built. Similar in overall appearance was a straight DC version. During production of the N80 and its DC counterpart, Soviet railways reclassified its electric fleet using the VL prefix to designate all domestically built electrics, which resulted in N60s becoming VL60s, the N80s becoming VL80s, and the DC versions becoming VL10s. By this time, the numbers had ceased to reflect axle weight.

The breakup of the Soviet Union resulted in the electric locomotive fleets being dispersed among the new state railways that once comprised the USSR while retaining their VL classifications, but often shorn of their large red stars that historically were used to decorate Soviet locomotives.

→ The Soviet N8 class electric was later reclassified as a VL8. Production of this rugged type began in 1955 and was phased out when the more modern VL80 was introduced in the early 1960s. Ukraine Railways' VL8-1301 rolls through Zaporizhzhya Male station on May 29, 2019, with a coal train. *Stephen Hirsch*

↘ Trailing view of a Ukraine Railways venerable VL8 class 3,000-volt DC electric, VL8-1582, in freight service at Zaporizhzhya Male station on May 29, 2019. *Stephen Hirsch*

↑↑ In Belarus, the 1960s-era VL80s still retain their Soviet red stars, such as this colorful example, VL80S-592, passing through Minsk Uschodi station on May 26, 2019. *Stephen Hirsch*

↑ Ukrainian Railways VL80-007 works west at L'viv with a heavily laden freight. Ukraine, like other portions of the former USSR, has exceptionally busy main lines where long freights pass every few minutes. *Brian Solomon*

→ The Soviet VL11 is a 3,000-volt DC locomotive with mechanical similarities to the VL80 family of AC electrics. This one rolls freight through the Ukrainian station Zaporizhzhya 1 on May 29, 2019. *Stephen Hirsch*

↓↓ A Belarus Railways VL80S leads a long freight through Minsk Uschodi station in May 2019. *Stephen Hirsch*

Budd Rail Diesel Cars

n the 1930s, Philadelphia-based Budd Company developed some of the earliest streamlined trains using its patented shot-welded stainless-steel designs characterized by lightweight construction and distinctive side fluting. These early streamliners helped Budd become a major supplier of passenger cars. In 1949, Budd introduced its Rail Diesel Car (commonly known as the RDC), a diesel-powered self-propelled car with a double-ended design that could be run as a single unit or together as multiple units making up longer trains. RDCs were powered by a pair of General Motors 6-110 Detroit diesels located beneath the car. Depending on the seating configuration, RDCs would be assigned to branch line, suburban work,

or long-distance runs. Budd offered several car types that emulated configurations of the old gas-electrics: the RDC-1 was configured entirely with passenger seats, the RDC-2 included a baggage compartment, the RDC-3 had baggage and Railway Post Office sections and a few coach seats, the RDC-4 was strictly a baggage car/RPO unit, and the RDC-9 was a cab-less variety intended to work as a center car between two other units. New Haven ordered a specialized lightweight streamlined train consisting of uniquely styled RDCs assigned to its *Roger Williams* service.

Like the old self-propelled gas-electrics and McKeen "wind-splitters," Budd's RDC provided railroads with a practical tool for cutting costs on lightly traveled routes and

↙ After an early January snow in 1987, Metro-North RDC No. 47, inherited from New Haven Railroad via Conrail, works old home rails in the Naugatuck Valley near Beacon Falls, Connecticut, on its way from Waterbury to Bridgeport. *Brian Solomon*

→ A Long Island Rail Road RDC glistens in the afternoon sun at the far eastern reaches of the railroad's passenger network at Montauk, New York, in May 1960. *Richard Jay Solomon*

suburban lines. It required fewer crew than locomotive-hauled trains and simplified operations because the RDCs didn't require turning facilities or run-around tracks (necessary for locomotive-hauled passenger trains). By the early 1960s, Budd had built 398 RDCs for North American railroads. These were most numerous in the northeastern states, where the largest fleets were bought by Boston & Maine, New York Central, and New Haven Railroads. Baltimore & Ohio was among the lines that ran them in long-distance service. Budd RDCs were less common in the west, although some railroads bought one or two RDCs for specific services. Southern Pacific bought one car to lower costs on lightly traveled Oakland–Sacramento trains. This lone RDC was later reassigned to SP's scenic Pacific Northwestern between Willits and Eureka, where it operated until 1971. Western Pacific, famous for its Budd streamlined *California Zephyr*, also bought a pair of RDCs to work as its *Zephyrette* on the 917-mile (1,475.8 km) Salt Lake City–Oakland run. They were virtually unknown in the American south, while the Canadian roads made good use of RDCs.

As railroads scaled back and eliminated many passenger routes in the late 1960s and 1970s, the number of RDC runs declined, yet some of the cars found continued work for Amtrak, which inherited most remaining US long-distance services in 1971. In the 1970s, commuter rail agencies and other passenger operators inherited fleets of RDCs, and these included Boston-based Massachusetts Bay Transportation Authority, New York–area Metro-North, Baltimore-based MARC, and Philadelphia-centered SEPTA, as well as Canada's passenger operator VIA Rail, created in the late 1970s to assume operations of Canadian National and Canadian Pacific.

Some cars operated for more than sixty years in regular service, and a few RDCs survive today on heritage railroads, including the Berkshire Scenic in Massachusetts,

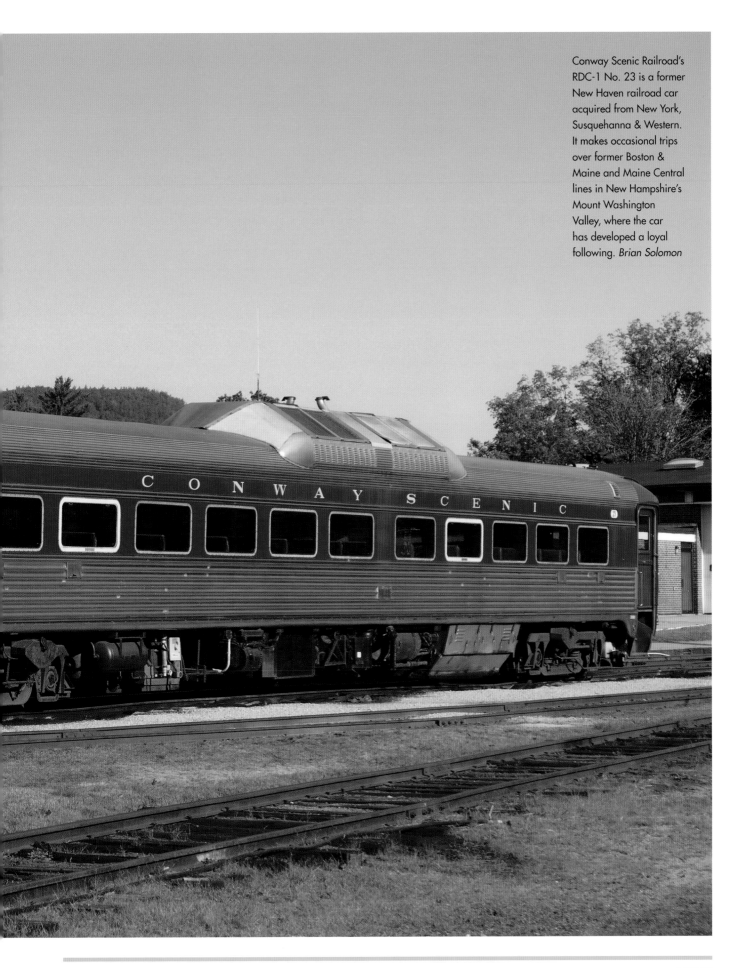

Conway Scenic Railroad's RDC-1 No. 23 is a former New Haven railroad car acquired from New York, Susquehanna & Western. It makes occasional trips over former Boston & Maine and Maine Central lines in New Hampshire's Mount Washington Valley, where the car has developed a loyal following. *Brian Solomon*

New Hampshire's Conway Scenic and Hobo Railroads, and Pennsylvania's Reading Northern, a freight carrier that runs classic railroad excursions.

In 1978, Budd reintroduced the self-propelled passenger car with its SPV-2000 model. This employed the Metro-shell body common to its Metroliner high-speed electric multiple units and Amtrak's Amfleet cars. Production was short-lived, so SPV-2000s were comparatively rare. However, thirteen of the cars were bought by the Connecticut Department of Transportation and were briefly operated by Amtrak on the New Haven–Hartford–Springfield shuttle, by Metro-North on suburban branch runs to Waterbury and Danbury, Connecticut, and by the former New York Central Hudson and Harlem lines in New York State.

↑ Reading & Northern 9166 is a rare operational example of an RDC-3 that featured a baggage compartment and area for a Railway Post Office. In July 2017, it was working near Grier City, Pennsylvania, on a return trip from Jim Thorpe to Reading Outer Station. *Brian Solomon*

← Reading & Northern Budd-built RDCs are ready for departure from Reading Outer Station, Pennsylvania. *Brian Solomon*

TALGO

n the 1930s and 1940s, Spanish inventor Alejandro Goicoechea Omar devised a novel form of railway passenger car. This is known by the TALGO acronym that combines the train's articulated design and the names of Omar and of his business partner, José Luis de Oriol y Urigüen: Tren Articulado Ligero Goicoechea-Oriol. In its original incarnation, Omar's TALGOs were distinguished from traditional passenger car design by employing fixed articulated sets of unusually short cars of lightweight tubular construction. These were semipermanently coupled, sharing common sets of wheel pairs that didn't use traditional trucks or continuous axles. Eliminating conventional four-wheel swiveling trucks reduced wheel wear and lowered tare weight while enabling a low center of gravity train with a low-clearance profile for comparatively high-speed operations.

In the late 1940s, TALGO contracted American Car and Foundry to construct several complete trainsets; three were built for Spanish National Railways, known by its initials RENFE (Red Nacional de los Ferrocarriles Españoles), and a demonstration train was made for an American tour in 1954 aimed at generating North American sales. At the time, US railways were exploring fast lightweight trains as a means to cut costs and boost sagging passenger revenue. Ultimately, Boston & Maine, New Haven, and Rock Island sampled the early TALGO trains but with marginal success. By contrast, RENFE made good use of its TALGOs and continued to buy advanced variations of the novel trainsets, which also were the first air-conditioned trains in Europe.

In the 1960s, RENFE introduced new fleets of TALGOs hauled by powerful German-designed Krauss-Maffei low-profile diesel-hydraulic locomotives, constructed under license in Spain. TALGO continued to innovate, producing some unusual and

→ A specially painted RENFE Class 269 electric leads a new Talgo VII set southward toward Malaga on broad gauge tracks. This tilting locomotive-hauled TALGO variation was introduced to intercity service by RENFE in 2000. RENFE has subsequently introduced several advanced TALGO trains that work new purpose-built AVE high-speed railway routes (built to the European standard gauge track width) and are capable of speeds up to 226 mph (365 km/hour). *Brian Solomon*

↙↓↓ Coming and going. On September 27, 2001, a RENFE Japanese-designed Class 269 electric whisks a classic 1960s-vintage TALGO III set southward at Villa Canas, Spain. The TALGO III was the third generation of TALGO equipment and was one among longest surviving types of intercity passenger trains in Europe until retired in 2009. *Brian Solomon*

← New Haven Railroad's short-lived TALGO-2 cars are pictured inside the railroad's Van Nest Shops in the Bronx, New York, on March 22, 1958. *Richard Jay Solomon*

→ ↘ An Amtrak TALGO at Portland, Oregon. In 1997 and 1998, Amtrak bought TALGO Series VI trainsets manufactured in Seattle, Washington, for its *Cascades* service between Eugene, Oregon, Seattle, and Vancouver, B. C. The large fiberglass fins were designed to offer a visual transition between the low-profile TALGO cars and the taller EMD F59PHI diesels bought to haul them. *Scott Lothes*

versatile variations of its passenger trains. The Spanish railways use broad gauge tracks, 5 feet 6 inches (167.6 cm) between inside rail faces (compared with the 4 feet 8.5 inches [143.5 cm] used by most lines in Europe and North America), and this difference greatly restricted through international rail travel; typically passengers had to change trains at the French–Spanish border. To overcome this impediment, TALGO designed a variable gauge axle to enable its trains to change gauge at slow speed, facilitating through service between Spain and France, which commenced in 1969. Gauge-changing TALGOs found additional routes with the opening of the Spanish high-speed line in 1992, known as Alta Velocidad Española, or AVE, which, unlike most of RENFE's network, was constructed to the European standard gauge. Gauge-changing TALGOs built for 125-mph (201.2 km/h) operation permit seamless through domestic passenger trains using both the AVE and broad gauge lines.

In 1980, the tilting TALGO pendular was introduced, employing a passive tilting system to minimize the effects of centrifugal forces on curves. TALGO again focused on the North American market in 1988 when it sent a pendular demonstrator on a tour.

Between 1994 and 1995, Amtrak imported twelve- and fourteen-car TALGO TP-200 trains for use in the Pacific Northwest and elsewhere. Later, Amtrak and Washington State cooperated to purchase TALGO Series VI trains for the *Cascades* regional service on the Eugene, Seattle, and Vancouver, British Columbia, corridor. Delivered in 1998, these feature a novel exterior design with large fiberglass fins to provide a better visual transition between EMD-built F59PHi diesels and the low-profile cars. Oregon bought a pair of thirteen-car Series 8 TALGO trains in 2013, while Wisconsin's plans for 110-mph (177 km/h) Series 8 TALGOs were foiled due to local political problems.

Hondekops

The Dutch national railway, Nederlandse Spoorwegen, known by its initials NS, was early to embrace wide-scale electrification. It used 1,500 direct current overhead catenary and adopted electric multiple units (EMU) for its standard passenger trains. Early Dutch passenger multiple units used a rounded front-end profile where the operator rode in a confined compartment at the point of the train, closely resembling the pattern established by the German high-speed diesel trains of the early 1930s. In the 1950s, NS adopted a distinctive EMU layout incorporating a cab design, which took its cues from modern streamlined diesel-electric locomotives as defined by the Electro-Motive E-unit

(and later adopted by the F-unit, as described on pages 90–95). This gave the operator an elevated position with a protective frontal nose section. Officially, these were designated Materieel '54, shortened to Mat '54; however, with its elongated bulbous front end and large slanted windows resembling the face of a sad hound, the trains were known colloquially in Dutch as "Hondekops." Between the mid-1950s and early 1960s, a variety of different car configurations or "Plans," including Plan F, G, M, P and Q types, were built by Dutch manufacturer Werkspoor at its shops in Utrecht. In addition, a three-piece diesel-electric multiple unit designated Plan U was also built by Werkspoor between 1960 and 1963. While the

Nederlandse Spoorwegen's multiple units are among the most distinctive looking trains in Europe. The Mat '64 Plan T and Plan V cars used a foreshortened nose section based on the earlier Mat '54 "Hondekop" cars. *Brian Solomon*

electric cars were painted green and cream, the diesel cars were given a bright red livery that earned them the name "Red Devils."

In the 1960s, an advanced train called the Materieel '64 evolved using much lighter construction to enable quicker acceleration and featuring a variety of modern innovations including improved door design. The front-end design resulted in a foreshortened pug-nose snout, and while still considered by many observers to be a variation of the Hondekop, they have been also described as Apekops, meaning "ape" or "monkey face." These were offered in two passenger varieties: the four-unit Plan T and the two-unit Plan V. Like the Mat '54 cars, the Plan T and early Plan V cars were constructed by Werkspoor while later cars were contracted to German supplier Talbot based at Aachen near the Dutch-German border. A single car variant, Plan mP, was designed as an electric postal delivery train ("mP" refers to "motor post"). Like the passenger variations, these could travel up to 87 mph (140 km/hour) in mainline service.

For decades, Nederlandse Spoorwegen's most common trains were the Mat '64 two-unit Plan V electric multiple units, such as these pictured at Eindhoven on May 22, 1996. The last of these was withdrawn from regular traffic in 2016, and a set was preserved at the Spoorwegmuseum in Utrecht. *Brian Solomon*

Postal service Mat '64 Plan mP single-car electric multiple units rest between runs at Den Haag (The Hague) in May 1996. These cars were retired from postal work in the late 1990s, with some cars being reassigned to infrastructure service. *Brian Solomon*

Swiss Re 4/4 and Re 6/6 Electrics

Switzerland boasts some of the world's busiest mountain main lines. The famous crossing of the Gotthard Pass was opened in 1882, featuring a superbly engineered line with a maximum gradient of just 2.7 percent and requiring several spiral tunnels to maintain an even gradient. The route was melded into the growing Swiss Federal Railway (SBB) network in 1909. The high availability of hydroelectric power combined with a dearth of domestic fossil fuels led Swiss railways to embrace wide-scale electrification relatively early, embracing the 15kV 16-2/3 Hz AC standard common in neighboring Austria and Germany.

Among the most impressive classic Swiss electric locomotives are SBB's Re 6/6, using the relatively unusual B-B-B wheel arrangement.

Introduced in 1972, these were part of a family of high-horsepower, high-adhesion Swiss electrics noted for achieving high power while keeping axle weight to 22 tons and placing the full weight of the locomotive on driving wheels. These use a welded body to reduce weight and were a contrast to early Swiss electrics characterized by heavy fabricated bodies and complex arrangements of powered and non-powered axles to evenly distribute locomotive weight.

The first of these earlier high-adhesion designs was the Berne-Loetschberg-Simplon Railway (BLS) Class Ae 4/4 electrics of 1944. These featured a B-B wheel arrangement powered by fully suspended single-phase alternating current motors. The precursor to SBB's Re 6/6 was the Ae 6/6 of 1952 that rode on a pair of C trucks, followed by the similarly designed four-axle Re 4/4 of 1964, which emerged as Switzerland's most numerous electric, built to accommodate a wide range of traffic and operating conditions. The Re 6/6, while retaining a degree of versatility, was intended for mountain operations on the Gotthard and Simplon lines. The design specifications enabled a single Re 6/6 to haul an 882 train over the Gotthard Pass at a sustained 50 mph (80.5 km/h). Increased train weights led SBB to operate combinations of Re 4/4 and Re 6/6s together to obtain the necessary power-to-weight ratio to keep trains moving over the road at desired speeds. These combinations are colloquially known as Re 10/10s and were common on Gotthard Pass freights until 2016, when the new 35.4-mile (57 km) Gotthard Base Tunnel opened, diverting most freight away from the steepest portion of the line via the older historic tunnel.

→ Nine of SBB's Re 4/4 electrics were painted in the burgundy-and-cream Trans Europ Express livery, such as 112552, seen leading an SBB intercity passenger train in 1988. *Denis McCabe*

↘ Wassen station on the Gotthard Pass is located near the mouth of a tunnel. A downhill freight is led by an Re 6/6 and Re 4/4 combination in April 2016, a few months before most freight was diverted to the new 35.4-mile (57 km) Gotthard Base Tunnel. *Brian Solomon*

↙ A detailed view of SBB Re 4/4 No. 11200 shows characteristic SBB stainless-steel raised numbers and the variety of manufacturers involved in its construction. *Brian Solomon*

← An Re 6/6 and Re 4/4 combination, known collectively as an Re 10/10, ascends the Gotthard Pass at Wassen, where line snakes through the village on three levels to gain elevation. *Brian Solomon*

↙ An uphill freight ascends the Gotthard Pass at Zgraggen, Switzerland. *Brian Solomon*

↗ Among the most famous locations on the Gotthard Pass are the Biaschina Spiral tunnels on the south slope. *Brian Solomon*

↘ An uphill freight crosses a bridge near Wassen on its way to the original Gotthard Tunnel. *Brian Solomon*

General Motors B121 Diesels

By the 1950s, General Motors Electro-Motive Division had virtually saturated the North American market and began working with overseas builders, licensing its designs and supplying key mechanical and electrical equipment. In 1947, Ireland's state-run transport company, Córas Iompair Éireann (known by its initials CIÉ), contacted General Motors regarding the development and purchase of its diesels. Intermittent dialogue over the next dozen years ultimately led to Ireland's purchase of fifteen GM model GL8 diesels that were among the builder's first exports to Europe from its domestic La Grange, Illinois, factory, establishing GM as Ireland's locomotive supplier for the next three decades.

The GL8 export model was a lightweight 875-horsepower endcab road diesel adapted in 1958 from an earlier G8 export type that itself was in effect an adaptation of the domestic model SW8 switcher. The first GL8s were built with A1A trucks (three-axle bogies with an unpowered center axle for weight distribution for light axle loading) for service on Taiwan's (China) railways. However, the standard GL8 arrangement, such as those that CIÉ bought, used a pair of lightweight powered twin-axle trucks. Most of the electrical equipment was GM standard components, including external hardware such as headlights. Comparatively squat locomotive cabs were necessary to conform with the low-clearance Irish loading gauge.

Ireland's first GM diesels were official designated as the B121 class, colloquially known as "Yanks," which initially distinguished them from the rest of the largely British-built diesel fleet. Although the B121s were intended for bi-directional operation, an unfortunate accident during early trials in Ireland led CIÉ to normally work the class cab first, restricting operations to the hood section leading. Because most of the network was still equipped with turntables for turning steam locomotives,

→ Although originally Class B121, in later years the 'B' was dropped and these locomotives were simply classified as 121s. Three of Irish Rail's fifteen Class 121s congregate at Dublin's North Wall freight yards in June 2000. Ordered without multiple-unit connections, the 121 diesels were later fitted for multiple working by the Inchicore shops in Dublin. *Brian Solomon*

↙ On January 13, 2003, a pair of 121s lead an empty cement train for reloading at the cement factory at Mungret, Co. Limerick. In later years, Irish Rail largely assigned its 121s to freight work. *Brian Solomon*

this didn't prove to be a significant operational hardship. Later orders for GM diesels were built with twin cabs. In the 1970s, CIÉ equipped the B121s with multiple-unit equipment that facilitated their operation with other GM types, thus minimizing complications from single locomotive operations and the necessity to turn engines at terminals.

The B121 Class GL8 was designed for freight and passenger services with mainline speeds in excess of 70 mph (112.7 km/h). The last two of the class were withdrawn from traffic in 2008 and preserved, with locomotive 134 presently under restoration for mainline excursions operated by the Railway Preservation Society of Ireland.

In July 1998, an engine driver leans from the cab of Irish Rail 123 to take the staff from the signalman at Dromod. The electric train staff was a simple, safe system that provided the train with absolute operating authority over the line. On this day, a pair of Class 121s were working a laden timber train from Sligo toward Dublin. *Brian Solomon*

Dv12 Diesel-Hydraulics

Finland's railway network had its origins when the nation was still under the control of Czarist Russia and as a result uses broad gauge tracks similar to those in Russia. Finnish State Railways (VR Group) is the national operator. Historically, VR's most common diesel locomotive was the Dv12 diesel-hydraulic, a type roughly equivalent to the GP9 in North America. This is a versatile, well-built type that could perform just about every job demanded of it.

The type had an unusually long production run that spanned two decades beginning in 1964 and were originally known as Class Sv12 (VR changed the classification in 1976). Interestingly, the Dv12s were not solely produced by one manufacturer, and production was divided between two different Finnish builders. Some were built by Lokomo and others by Valmet. There are three varieties, with minor differences between them. Each is grouped by their number series: 2501–2568, 2601–2664, and 2701–2760. Last built were the 2700 series. All of the Dv12s were originally powered by a Tampella sixteen-cylinder diesel engine, a French design assembled under license in Finland.

Among their unusual features that gave them added versatility was the ability to easily switch between two gear ratios, which allowed them to work in both freight and passenger service. In addition, VR routinely used them as switching engines, while operating them in multiple sets of two or three in road service. Despite their age, more than one hundred of the class remained active as late as 2015. In recent years, their numbers have diminished, a result of Finnish railways extending electrification on many principal routes. The classic Dv12 red-and-cream paint livery has given way to VR's modern white and lime, with many locomotives wearing a patched livery during the transition.

↙ For many years, Oulu was the end of mainline electrification, and many trains switched between diesel and electric locomotives at this point. On a summer 2001 evening, Dv12 2627 goes for a spin on the turntable at the Oulu roundhouse. At right is a Dr16 AC-traction diesel. *Brian Solomon*

→ VR Group Dv12 2617 arrives at Oulu, Finland, with a short freight. These versatile diesel-hydraulics were routinely assigned to both freight and passenger duties. *Brian Solomon*

← Looking down on the short hood of Dv12 2624 wearing the classic VR paint scheme. These were full bi-directional locomotives built to the road-switcher configuration. *Brian Solomon*

→ A close-up of Dv12 2648's builder's plate. *Brian Solomon*

↘ In recent years, VR Group has been repainting surviving Dv12s in the railway's current white-and-lime-green livery, which makes for a modern contrast to the traditional paint schemes this class has labored in for more than five decades. *Brian Solomon*

Series 0 Trains

The original Japanese Shinkansen was the 343-mile (552 km) New Tokaido Line, an all-new railway that opened in 1964 to improve and increase rail service on Japan's busiest route between Tokyo and Osaka, its two largest cities. Planning began in the 1950s, and the Shinkansen's 1964 debut coincided with the Tokyo Olympic Games, which opened the world's eyes to Japan's outstanding achievements.

Shinkansen means "new main line," and unlike most of the Japanese network that is built to narrow gauge standards, this all-new high-speed railway was constructed to the European/American standard track gauge—4 feet 8.5 inches (143.5 cm) between the rails to allow for greater stability at speed and larger trains with greater carrying capacity.

The New Tokaido Line was engineered roughly parallel to the older Tokaido Line but followed an entirely new alignment, which significantly shortened the distance by reducing curvature, eliminating all grade-level road crossings, and enabling average speeds faster than 125 mph (201.2 km/h). The new railway ran southwest from Tokyo, passing Japan's iconic volcanic cone of Mt. Fuji and intermediate stations, including the ancient capital of Kyoto.

Shinkansen service was inaugurated with the famous bullet-nose blue-and-white Series 0 high-speed trains. Despite their stunning, sleek appearance, these fixed-consist multiple-unit trains didn't employ revolutionary technology. These were essentially highly refined adaptations of electric multiple units used by American electrified commuter lines (Japanese National Railway's chief engineer apprenticed before World War II on New Haven Railroad). Technically, Japan's Series 0 trains were near cousins to Pennsylvania Railroad's classic owl-eyed MP54s used in

↙ At Nishi-Akashi, Japan, an all-stops *Kodama* service operated by a vintage Series 0 "Bullet Train" is overtaken by a faster express service worked by a more modern Series 100 train. Today, faster, sleeker trains than the rolling antiques of the 1960s and 1970s ply the Shinkansen. *Brian Solomon*

→ One of the iconic Series 0 "Bullet Trains" recedes into the tunnel at Kobe, Japan. This was working an all-stops *Kodama* local run toward Tokyo. Newer, faster trains have been assigned the premier Shinkansen trips. *Brian Solomon*

New York and Philadelphia suburban service. However, to reach and sustain speeds of up to 135 mph (217.3 km/h), Japan's Series 0 trains used an exceptionally high power-to-weight ratio and an aerodynamic design. The Shinkansen and its "Bullet Trains" proved an immediate success, and during its first three years, the new railway carried 100 million passengers. This success spurred extensions and several new Shinkansen routes.

Series 0 trains remained in production for twenty-three years, resulting in more than 3,200 individual cars working 12- or 16-car trainsets. In later years, the iconic trains were bumped to less glamorous runs, as newer and more advanced train designs were built for the fastest and most prestigious runs. The privatization of Japan National Railways (JNR) in 1987 split Shinkansen operations among three different companies and spurred development of new generations of high-speed trains. These embraced new technologies and futuristic streamlining both to enable speeds up to 200 mph (321.9 km/h) and to minimize wind roar. The most modern Shinkansen trains now include the striking E3 and E5 Series with their exceptionally long tapered ends.

← A Japanese classic: in April 1997, a vintage Series 0 Shinkansen high-speed train glides away from the station in Kobe. *Brian Solomon*

↙ Interior view of a Series 100 Shinkansen train's first-class section. *Brian Solomon*

→ A Series 700 Shinkansen train meets a JR West Series 500 train near Tokyo. The Series 700 trains were introduced in 1999. *Scott Lothes*

↘ A JR West Series 500 train working as a Nozomi Express catches a glint of light in Tokyo in 2007. The more modern Shinkansen trains employ advanced wind-resistant designs to keep noise levels to acceptable levels. *Scott Lothes*

M62 Diesels

The mass-produced M62 carbody-style diesel-electric was developed by the Soviet Union in the mid-1960s based on previous Soviet diesel designs, including road-switcher models derived from American Locomotive Works diesels exported from the United States to the USSR in the mid-1940s. The 2,000-horsepower M62 was built at Lugansk's October Revolution Works and powered by a two-cycle, twelve-cylinder Kolomna diesel originally designed for marine applications. These routinely worked both freight and passenger services. Nearly 2,500 double-cab single M62s were built between 1965 and 1980, many for the USSR's European satellites, with Hungary being the first recipient; there, they were colloquially known as "Sergeis," in reference to their Soviet origin. M62s could be found across Eastern Europe,

including Poland, Czechoslovakia, and East Germany, as well as in the USSR's various component nations. They were also exported to Cuba and North Korea.

Beginning in 1976, a modified two-piece, semi-permanently coupled variation designated 2M62 was introduced and built in large numbers; this type was later advanced into model 2M62U, while a three-piece 3M62U found limited applications in Russia and Kazakhstan. After the breakup of the USSR, M62 variants continued to work for newly created state railways across the former Soviet sphere. A few M62s were reengineered with MTU and Caterpillar diesels, while Rail World's Rail Polska private freight operator began rebuilding M62s with EMD 645F diesels. The prototype M62-1 has been preserved in St. Petersburg.

A 2M62 leads an eastward container train at Tapa, Estonia, in July 2002. The semi-permanently coupled 2M62 was essentially a pair of single-cab M62s back to back. *Brian Solomon*

← In the mid-2000s, Rail World's Rail Polska rebuilt Soviet M62s with EMD 645 diesels to work freights in Poland. *Brian Solomon*

↙ A sharply painted Korean State Railway 700 class (M62) locomotive, No. 705, leads a freight under wire on the North Korean main line between Wonsan and Pyongyang. *Stephen Hirsch*

Class Rc Electrics

After World War II, Sweden, which had been a pioneer in electrified railway operations, accelerated its investment in electrification and continued to push electric locomotive technology. In the 1960s, the Swedish firm Allmänna Svenska Elektriska Aktiebolag (ASEA) developed the Class Rc electric for dual service. The first Rcs entered service on Statens Järnvägar (the Swedish national railway) in 1967. Significantly, these electrics pioneered commercially thyristor motor control on a locomotive. Thyristors are high-voltage semiconductors that were used in place of traditional electro-mechanical or pneumatic motor controls. These offered stepless traction motor regulation that permits maximum motor output while controlling wheel slip for superior adhesion. The end result was greater efficiency and much higher tractive effort. Swedish Rc models have included Rc-1s to Rc-6s, plus a heavy freight locomotive with slow speed

gearing designated Class Rm. Another variation was the Rc-7, a short-lived re-geared Rc-6. Rc adaptations have worked in more than a half dozen nations, including Austria, Bulgaria, Croatia, Iran, Norway, Romania, and Serbia, making it one of the most widely sold types of the 1970s and 1980s.

In the 1970s, Amtrak was facing the need to replace its aging fleet of former Pennsylvania Railroad GG1 electrics and was dissatisfied with its experience with GE-built E60C electrics in high-speed service and imported electrics from Sweden and France for comparative testing. It found that the ASEA Rc-4 best suited its application, so this type was adapted to the model AEM-7 for service on its electrified Northeast Corridor. These were assembled under license by EMD at La Grange, Illinois. They resembled the Swedish prototype but featured a much tougher body shell, necessary for American safety requirements, and were significantly more powerful in order to

Tågkompaniet was the operator of local passenger trains running on the line north of the Arctic Circle over the Malmbanan that connects ore mines in northern Sweden with the port at Narvik, Norway. Here, an Rc-6 leads a local from Narvik southward along the Arctic Lake known in Swedish as the Torne Trask. *Brian Solomon*

↑ On August 1, 1986, Amtrak AEM-7 926 races along the Northeast Corridor at Linden, New Jersey, with a Philadelphia-bound *Clocker. Brian Solomon*

← SEPTA AEM-7 2308 rests between runs at West Trenton, New Jersey. Philadelphia-based SEPTA's eight AEM-7s were delivered in 1987. *Brian Solomon*

reach higher speeds. Initially, Amtrak ordered forty-seven of the type but later bought additional locomotives. In addition, NJ Transit and SEPTA bought AEM-7s and derivative types. In their final years, Amtrak modernized the AEM-7s with a three-phase AC-traction system, redesignating the locomotives as AEM-7 ACs. As the new Siemens ACS-64s entered service beginning in 2014, Amtrak began withdrawing its AEM-7 fleet. SEPTA sidelined its AEM-7s and lone ALP44 (an AEM-7 derivative) in 2019.

Statens Järnvägar (Swedish State Railway) Rc-6 1329 is ready to depart Lulea with an overnight sleeping car train bound for Stockholm. Passengers will need to keep the blinds drawn if they plan to get any sleep, because in July, it never gets completely dark in the far north. *Brian Solomon*

Class 103 Electrics

Germany's iconic Class 103 electric is among the most recognizable European locomotive. A new electric was needed to fulfill Deutsche Bundesbahn's (Germany Federal Railway) renewed interest in developing high-speed electric services in the 1960s instead of using lightweight diesel trains as it had with its *Flying Hamburger* trains in the 1930s (see page 76). Four prototype electrics, designated E03, were built in 1965 and designed for 125-mph (201.2 km/h) operation and were capable of unusually rapid acceleration. The body was designed with the aid of wind-tunnel testing, used a five-section welded streamlined body, and rode on C-C trucks, which was a novelty in Europe for high-speed electric design but allowed for smaller traction motors with better weight distribution. Regular production spanned from 1970 to 1974 and consisted of another 145 units.

A restored Class 103 235-8 leads a preserved *Rhein Gold* trainset at Föhren on the line from Koblenz to Trier as part of celebrations in April 2010 commemorating 175 years of German railways. *Brian Solomon*

Most of the Class 103s could deliver continuous output of approximately 8,000 horsepower, with bursts of 12,000 horsepower to accelerate trains quickly. This was key to the locomotive operating demands, because getting up to top speed quickly was necessary to maintain schedules, as many German intercity passenger trains stopped often.

In their classic form, the 103s were painted in an attractive burgundy-and-cream livery to match Trans Europ Express (TEE) luxury international consists. The Class 103 remained as Germany's premier passenger electric for three decades until it was supplanted in the late 1990s by both Deutsche Bahn's (the railway was renamed after German reunification) modern superfast InterCity Express (ICE) trains and modern Class 101 electrics working conventional passenger train consists. In 1998, the Class 103 enjoyed a brief resurgence when a tragic accident with an ICE-2 train resulted in the temporary withdrawal of the newer trains for reengineering, while numerous 103s were restored to traffic. Most of the class were withdrawn by 2001; however, more than a dozen 103 locomotives have preserved assignments in vintage liveries, and several continue to work regular assignments while making appearances on heritage excursions.

← On a warm May 1996 evening, DB Class 103 236-6, still wearing its classic burgundy-and-cream livery, leads an InterCity train northward through the Rhein Valley at Namedy. *Brian Solomon*

↗ Although supplanted by more modern locomotives and ICE high-speed trains, a few Class 103s have been preserved and continue to be assigned to revenue Deutsche Bahn passenger trains. In May 2011, Class 103 245-7 was seen at the Munich main station. *Brian Solomon*

→ Dressed in Deutsche Bahn's Spartan red and white, a Class 103 races an InterCity train northward through Bonn in August 1998. *Brian Solomon*

Class ET22 Electrics

After World War II, Polish State Railways (PKP) began electrifying primary routes using a 3,000-volt direct current overhead system to take advantage of large deposits of domestic coal for generating stations. When the electrification program accelerated in the 1970s, one of PKP's standard electric locomotives was its Class ET22, a full-carbody type riding on six powered axles. These were mass-produced domestically by Pafawag using Dolmel electrical gear. The body is a traditional boxcab with semi-streamlined styling. The locomotive is powered by six series-wound traction motors—one per axle—that produce up to 4,020 horsepower for traction. Although intended primarily as a freight-service locomotive, the ET22 has often been assigned to passenger work, with long sleeping car trains connecting points in Poland with cities across the former Eastern Bloc, including Russia and Ukraine.

PKP's mainlines are well built but with relatively conservative top speeds, which keep both freight and passenger trains traveling at comparable speeds. Among the most common assignments for the ET22 is moving heavy coal trains and other long freights. Nearly 1,200 of the type have been built since 1969, making this one of the largest unified fleets of electric locomotives in Europe, much larger than any single class of electric locomotive in North America. In addition to those constructed for PKP, twenty-three locomotives of a derivative design, known as Class E-1000s, were built for export to Morocco. Traditionally, PKP locomotives were painted two-tone green with yellow fronts, featuring a prominent pair of low front headlights.

↓ As one of PKP's standard freight locomotives, the ET22 class has often worked coal trains, such as this one approaching Jaworzyna. *Brian Solomon*

→ An ET22 passes a classic German-style mechanical semaphore heading westbound at Gniezno, Poland, in May 2000. *Brian Solomon*

Electro-Motive DASH 2 Diesels

By 1972, American diesel builder General Motors' Electro-Motive Division dominated the North American locomotive market. Despite its market supremacy, the company continued to innovate and implement improvements to its already excellent locomotive line. Previous improvements were reflected by changes to individual model designations. By contrast, its 1972 electrical and mechanical improvements were introduced across its locomotive line, so to avoid confusion, it continued to use its established model designations, denoting the change by adding *-2* at the end of the model number. The -2 models retained the same configurations and output as the older models they replaced.

The -2 models represented a high-water mark in electro-motive production, and the -2 models set standards for excellent performance and high reliability; these models were some of the best-selling North American diesel locomotives of the twentieth century. Among the new -2 models were the 2,000-horsepower GP38-2 and 3,000-horsepower GP40-2 four-motor four-axle models, and the six-motor SD40-2. Today, numerous examples of these diesels can still be found at work, with third-party rebuilds commanding high prices in the used locomotive market.

GM adapted the equipment arrangement of the freight service GP40-2 into a 2 variant called F40PH for passenger service. This was protected by a full-width cowling and initially offered head-end power from the 645 diesel prime-mover. In 1976, Amtrak was the first customer, ultimately acquiring more than two hundred; these engines became the face of its passenger trains for nearly two decades. The F40PH hauled trains across its network, from the Northeast Corridor expresses in non-electrified territory between New Haven, Connecticut, and Boston, to its popular long-distance trains, including the

A quartet of Burlington Northern SD40-2s roar upgrade on the former Great Northern line over Montana's Marias Pass. EMD's SD40-2 represented a high-water mark for reliability and durability; more than forty-five years after the model was introduced, many still work American lines. *Brian Solomon*

↑ Santa Fe was one of several railroads that bought the 3,600-horsepower SD45-2, which, while a more reliable locomotive, was not as popular as EMD's pre-2 SD45 built from 1966 to 1971. An eastward Santa Fe piggyback train crosses the Mojave Desert at Edwards, California. *Brian Solomon*

← In July 2017, a pair of Wisconsin & Southern SD40-2s lead a freight across Lake Wisconsin near Merrimac. *Brian Solomon*

Chicago–Oakland *California Zephyr* and the Chicago–Seattle *Empire Builder*.

Commuter rail agencies followed Amtrak's lead, with Boston-based Massachusetts Bay Transportation Authority being among the first to buy new F40PHs to replace aging postwar commuter diesels. Variations of the F40PH include the F40PH-2 and F40PHM-2 (featuring an unusual flush-front cab that was only ordered by Chicago's Metra). Although Amtrak retired most of its F40PH fleet as it acquired new GM-built *Genesis* models, many of the type survive in commuter service, notably on Chicago's Metra and California's San Francisco–based Cal-Train, plus on Canada's VIA Rail long-distance services.

← Massachusetts Central is a 26-mile (41.8 km) short line serving its namesake region using a pair of former Penn Central 2,000-horsepower GP38-2s painted to resemble Boston & Maine locomotives from the mid-1970s. *Brian Solomon*

↙ A New Jersey Transit F40PH leads a Raritan Valley push-pull train at Hunter Tower in Newark, New Jersey, in August 1986. *Brian Solomon*

↗ Cal-Train F40PH-2s are lined up at 4th & Townsend Street in San Francisco. Cal-Train provides an intensive diesel push-pull suburban service on the former Southern Pacific Peninsula Line between San Francisco, San Jose, and Gilroy, California. *Brian Solomon*

→ The Chicago-based Metra operates a large fleet of EMD F40PHs in regional suburban service. In February 2003, four of the smartly painted units are running toward the yard near the A2 tower in preparation for the evening rush hour. *Brian Solomon*

British Rail HST (High Speed Train)

Britain's High Speed Train (HST) was among the positive developments emerging from the austerity and drastic cuts of the "Beeching Era." In the 1960s, the state-run railway company (known as British Railways until 1965, when it was rebranded British Rail) had undergone forced contraction in the face of rapidly increasing automobile ownership and declining passenger ridership. Draconian cuts were implemented under the administration of Dr. Richard Beeching, who, following political instructions, had conducted clinical studies on the state of traffic on the network. Yet Beeching and his successors also desired to make investments where it would benefit the greatest number of people. In contrast to Japan and France, which were developing new high-speed railway networks that emphasized long sections of purpose-built high-speed tracks, Britain needed a modern *train* that could dramatically improve domestic intercity services and boost ridership without requiring expensive new infrastructure.

So in the 1970s, BR invested in two parallel train designs. The Advanced Passenger Train, or APT, employed a sophisticated tilting suspension system and advanced propulsion for top speeds up to 160 mph (257.5 km/h) on existing tracks. By contrast, the HST was developed by refining established diesel-electric propulsion without the added complexity of active tilting suspension, while keeping top operating speeds to a more conservative 125 mph (201.2 km/h). Among the advantages of the HST is rapid acceleration without the need for electrification and the ability to use existing main lines with few changes except for some adjustments to signaling. Meanwhile, flaws in the APT design led BR to abandon that project and instead favor the HST for widespread applications.

HST was conceived in 1970, a prototype began testing in 1972, and revenue trains

↙ To celebrate its iconic HSTs operator, LNER repainted a seven-coach set with the original BR livery and operated it on a four-day farewell rail tour to various East Coast destinations, including Inverness, Aberdeen, Leeds, and London. It is pictured passing the Bow of Fife in Scotland. *Finn O'Neill*

→ Britain's Cornish Main Line features numerous curves and viaducts. On May 10, 2018, a Great Western Railway "Castle Class" HST crosses the Largin Viaduct on its way from Exeter St Davids to Penzance on the historic Great Western Railway route. Today's GWR is one of several modern operating companies in the UK that has adopted the historic name of the route that it operates on. While this pays tribute to history, it can be confusing because modern operators have no corporate connections to the historic companies from before British railway nationalization in 1948. *Finn O'Neill*

entered service in 1975 and 1976. Each HST set featured purpose-built, wedge-shaped Class 43 diesel-electric locomotives at each end of a semipermanently arranged train of seven Mark 3 passenger cars (consists were later expanded to eight cars). As built, the locomotives used twelve-cylinder Paxman Valenta diesels rated at 2,250 horsepower, giving the train 4,500 horsepower, which supplied ample power to accelerate from a dead stop to maximum line speed in under six minutes.

To enable the HST to safely operate at much faster speeds on the existing network, trains have to comfortably slow and stop in just 6,600 feet (167.6 m) from their 125 mph (201.2 km/h) cruising speed as defined by BR signaling. The HST was the first British wide-scale adaption of electropneumatic disc braking, a system applied to all wheels, along with additional advanced braking, to prevent wheels from sliding.

Key to the success of HST technology was the comfortable Mk3 passenger car, employing a modern monocoque body featuring a very strong and lightweight design.

The HST was marketed as the Intercity 125 and was gradually introduced on a variety of intercity routes, beginning with service on the former Great Western Railway westward from London's Paddington Station. The new service was an immediate success, setting important service parameters and precedents that all passenger operators could learn from. Not only were HSTs newer and cleaner, operating faster and more comfortably than older trains, but they also benefited from more frequent schedules throughout the day and were priced the same as existing trains, thus giving rail passengers a much better service at no additional cost.

HST saved BR's intercity services; in the 1970s, it offered comparable station-to-station journey times as Japan's superfast Shinkansen and established records as the world's fastest diesel power train. By 1982, BR had ninety-five HST sets in service across its national network.

While BR was privatized in the 1990s, many of the HSTs survived for another quarter century, serving a host of private operators. Many trains were rebuilt and refurbished, with modern MTU diesels replacing the original Paxman Valenta engines. In 2019, electrification and modern trains finally saw the end of most regular HST operations in the United Kingdom.

↖ Most of the Great Western Railway long-distance services between London, Devon, and Cornwall operate along the Berks & Hants line between Reading and Taunton. In this view, an HST wearing the obsolete First Great Western blue livery accelerates away from Westbury on its way from London to Exeter St Davids on November 8, 2017. *Finn O'Neill*

↑ In the United Kingdom, the modern train operator London North Eastern Railway borrowed its name and initials from the similarly named historic pre-British Railways company London & North Eastern Railway. The modern operator assumed the East Coast Main Line franchise from Virgin Trains East Coast in 2018. In the fading winter light on December 2, 2019, LNER-lettered HST power car 43277 whisks a Leeds-to-London train running on the East Coast Main Line at Sandy in Bedfordshire just two weeks before LNER concluded its regular operation of HSTs. *Finn O'Neill*

← On May 11, 2010, First Great Western HSTs were lined up beneath the Victorian-era train sheds at London's Paddington Station. *Brian Solomon*

↑ LNER's HST in heritage paint skirts the shores of the Firth of Forth near Edinburgh in Scotland on December 18, 2019, shortly before the end of LNER's HST services. *Finn O'Neill*

↗ An East Coast HST pauses for passengers at Leeds in July 2014. East Coast operated select intercity services in the United Kingdom for about six years. Despite their age, the versatility, speed, and comfort of the Class HST sets have resulted in many of the 1970s-era trains surviving for more than forty years in daily service. *Brian Solomon*

↑ For a generation of British railway travelers, the HST represented a quick and comfortable alternative to traditional trains. An HST Class 43 power car wearing the late BR-era Intercity livery leads a train on the Great Western route in London. *Brian Solomon*

Class 120 Electrics

Among the most influential locomotives in the late twentieth century was the German Class 120. This pioneered the modern three-phase AC propulsion system pushing the limits of four-axle electric locomotive performance. Prototypes were ordered in 1976 and delivered in 1979; however, several years passed before the Class 120 entered production. In the mid-1980s, sixty production Class 120 electrics were built by several manufacturers for Deutsche Bahn (German Railways). The Class 120's innovative polyphase AC electrical system driving asynchronous traction motors developed by Brown Boveri & Cie. aimed to enable a single high-output B-B type locomotive to work both 125-mph (201.2 km/h) passenger trains and heavy freights, thus eliminating the necessity of maintaining separate pools for freight and passenger electrics.

Although the Class 120 didn't achieve its anticipated success in Germany and ultimately was primarily applied as a passenger locomotive, the type was later adapted in the 1990s for applications on other European national railways. More significantly, its three-phase traction set important precedents for further development. Today, three-phase AC traction is widely used for railway propulsion around the world, having been adapted for high-speed and heavy-haul applications. While in decline as of 2020, some of DB's Class 120s remain in German intercity passenger services.

Beginning in 1992, Spanish State Railway RENFE ordered Siemens Euro Sprinter electrics Class 252s that were derived from the Class 120 and shared a similar appearance. Spain initially acquired standard gauge versions of the Class 252 to haul TALGO tilting trains on its new high-speed lines and later acquired fleets of similar locomotives built for its broad gauge system tracks (5 feet 6 inches [167.6 cm] between inside rail faces). Similar electrics were bought by Compoios de Portugal, Portugal's national railway (also a broad gauge system), and Greece's Hellenic Railways, which use standard 4-foot-8.5-inch (143.5 cm) gauge tracks.

↙ The German Class 120 features a distinctive angular profile and was designed for moderately fast running, up to 125 mph (201.2 km/h) in passenger service. *Brian Solomon*

→ In recent years, Class 120s were becoming relatively scarce in intercity passenger service, yet on September 18, 2019, 120-146-6 was still at work leading a northward IC train through the iconic twin medieval towers at Oberwesel, Germany. *Brian Solomon*

↖ Portugal's national railway, Compoios de Portugal, has a fleet of Class 5600 electrics derived from Germany's Class 120. These have worked both freight and passenger trains. On March 29, 2019, a 5600 Series whisks a Lisbon-bound intercity train through the Porto suburb of Valadares. *Brian Solomon*

← Compoios de Portugal's 5617-4 bears the names of its German builders Siemens and Krauss-Maffei on its side as it glides through Porto's Campanhã station. *Brian Solomon*

↑ The Spanish State Railway, known by the initials RENFE, bought two fleets of Class 252 electrics based on the German Class 120. One fleet was built to work Iberian 5-foot-6-inch (167.6 cm) broad gauge lines; the other fleet, as exemplified here, work Spain's high-speed lines that were engineered to the European standard gauge (4 feet 8.5 inches [143.5 cm]). *Brian Solomon*

→ DB 120-114-4 catches the evening sun in August 1998 working at the back of a push-pull intercity train at the Bonn main station. *Brian Solomon*

TGV (Trains à Grande Vitesse)

n the 1960s, the French national railway, Société Nationale des Chemin de Fer (SNCF), aimed to dramatically improve its long-distance passenger service by augmenting existing routes with new purpose-built high-speed railway lines (known in French as Lignes à Grande Vitesse, or LGV) while engineering new high-speed trains (called Trains à Grande Vitesse, or TGV) to make the most of the new routes. These trains were overhead electric, semipermanently coupled articulated double-ended sets with high-output power cars at both ends. Significantly, while the new TGV equipment was able to reach the fastest speeds on the new routes, these were also designed for full compatibility with the existing network (albeit at conventional speeds), a feature that was key to lowering costs while enabling the new trains great route availability.

TGV train technology has been at the forefront of high-speed train development. Distinguishing between TGV high-speed train technology and SNCF's TGV branding often confuses observers. TGV branding is used by SNCF for its internal and internationally operated high-speed rail service, which use TGV trainsets. However, numerous high-speed systems around the world have adopted TGV technology but use distinctively branded trains, such as Spain's AVE network, the Paris/Brussels–London *Eurostar* services, and Paris–Brussels–Köln/Amsterdam *Thalys* services. In the United States, Amtrak's original *Acela Express* high-speed trains benefited from TGV technology, including Alstom's advanced propulsion system.

SNCF debuted its TGV on the heavily traveled Paris–Lyon corridor in 1981. The immediate success of this high-speed service propelled a French railway renaissance, and, in the last forty years, a national network of high-speed lines has been constructed radiating from Paris, including a variety of international routes, to shorten running times and improve rail travel. Today's TGV-branded network connects Paris with numerous cities across France and to

→ SNCF's original TGV fleet was the Paris Sud Est trains built for the Paris–Lyon run and featured the early 1980s orange, brown, and cream livery. *Denis McCabe*

↘ A TGV-PSE glides along at high speed in 1991. These trains initially had a top speed of just 161.6 mph (260.1 km/h). Modern SNCF TGV trains now operate at more than 200 mph (321.9 km/h). *Denis McCabe*

↙ SNCF has operating arrangements with many railways across Europe. This TGV Duplex—a bi-level train—is seen blitzing Rastatt, Germany, en route to Stuttgart. *Brian Solomon*

← TGV direct trains from greater Paris to Switzerland are marketed as TGV Lyria. A specially painted train makes its station stop at Paris Charles de Gaulle Airport. *Brian Solomon*

↙ At left is a specially painted multi-voltage TGV set for the Paris–Milan run; at right is one of the more modern TGV Duplex trains. *Brian Solomon*

→ First class on a TGV is comfortable and spacious yet not especially luxurious, but it represents a step above commercial airline travel. *Brian Solomon*

fifteen nations across Europe, including Spain, Germany, Switzerland, Italy, and the United Kingdom. In addition, a circumferential route bypasses central Paris to serve Paris Charles de Gaulle Airport station as a secondary hub, which allows direct transfer from international flights to the SNCF high-speed network.

The initial TGV services operated at maximum speed of 161.6 mph (260.1 km/h) in revenue service. Over the years, special test runs have pushed record speed to new highs (over 310 mph [499 km/h]), while revenue speeds have been gradually increased in conjunction with new highly engineered LGV routes and more refined trains so that today some services operate at top speed of nearly 200 mph (321.9 km/h) in revenue service.

Pendolinos

Tilting trains are simply a means of accelerating train schedules without a massive investment in new infrastructure. Low center of gravity and tilting train bodies allow for significantly faster train speeds through curves while minimizing the uncomfortable effects of centrifugal forces on passengers. There are two basic types of tilting train mechanisms: active and passive. Active tilting mechanisms use some mechanical or electrical apparatus to force a tilt to counteract centrifugal forces; these systems have been employed by the British experimental APT (advanced passenger train), the Canadian-designed LRC, and its derivatives, such as Amtrak's *Acela Express* high-speed trains. Passive tilting trains are designed to employ natural forces for the tilting effect. Italy's Fiat Ferroviaria–designed Pendolinos, employing a pendular mechanism, are among the most widely implemented passive tilting trains in Europe.

Italy's original fleet of Pendolinos was nine-car, double-ended electric multiple-unit Class ETR450s built in the 1980s. These drew power from 3,000-volt DC overhead electrification, operating up to 155 mph (249.5 km/h) on Italy's purpose-built, high-speed Direttissima routes, and were also capable of working conventional routes at speeds up to 30 percent faster than traditional locomotive-hauled trains.

Fiat Ferroviaria advanced its Pendolino technology in the 1990s, introducing the angular-shaped ETR 460. The Pendolino's tilt system offers a smooth ride on sinuous track while effects of the tilt are subtle and barely detectable at speed. Variations of the Pendolino design have been adapted for use in many European railways, including the Czech Republic, Finland, Germany, Portugal, Russia, Slovenia, Slovakia, and Switzerland. Britain's Virgin trains ordered a specially styled variation for high-speed services in the United

↙ An Italian State Railways ETR460 Pendolino tilts as it blitzes the station at Framura, which is located on the side of a cliff overlooking the Mediterranean Sea. *Brian Solomon*

→ Britain's Virgin Trains bought Alstom-built Fiat Pendolinos for service on the West Coast Main Line. The first of these distinctively styled trains entered service in 2002. The red, yellow, and silver painted Class 390 Pendolino makes for a stark contrast with the Victorian-era train shed at Manchester Piccadilly station in August 2014. *Brian Solomon*

← Alstom's ETR610 high-speed tilting trains operate in EuroCity services via Gotthard and Simplon Passes in the Swiss Alps. This Italian State Railways Trenitalia Pendolino was seen ascending the old Gotthard route near Zgraggen, Switzerland, in April 2016.
Brian Solomon

→ Trenitalia ETR610 glides through Flüelen, Switzerland, on the famous Gotthard route in April 2017. These sleek, modern trains were built by Alstom at Savigliano, Italy, in a factory famous for three decades of Pendolino construction.
Brian Solomon

↘ The Czech Republic's national railway, České Dráhy (CD), operates a small fleet of Italian-designed Pendolinos in *SuperCity* service between Prague and Ostrava. On a misty October 2016 afternoon, one of CD's Pendolinos tilts through a curve at Drahotuse on its run toward the Czech capital. *Brian Solomon*

Kingdom, running north from London Euston Station over the electrified West Coast Main Line. The Spanish national railway RENFE ordered non-tilting variations of the Pendolino.

In 2000, Alstom acquired Fiat Ferroviaria and continued to advance tilting train design.

The sleek, modern Alstom Avelia Pendolino represents the fourth generation of this proven tilting train system, and it is among the firm's latest offerings, boasting that more than three hundred trainsets have been built with Pendolino tilting technology.

General Electric Genesis

General Electric's Genesis diesels were an early 1990s development stemming from Amtrak's need to replace its fleets of high-mileage 1970s-era diesels. In the 1970s and 1980s, North American passenger locomotives were largely adaptations of standard freight types. For its next generation, Amtrak desired a purpose-built, lightweight, state-of-the-art passenger locomotive embodying the latest European designs. Both of the primary diesel builders bid on the new project, which Amtrak designated the AMD-103 (Amtrak Diesel, 103 mph [165.8 km/h]). GE won the contract, and "Genesis" was the result of a company-naming contest.

Ultimately, GE developed three distinct Genesis models, all of which use a distinctively styled integral monocoque body shell and fabricated trucks instead of a bottom-supported locomotive platform and cast trucks typical of its freight locomotives. GE worked with Krupp for the body design. To allow the Genesis models to access lines with restrictive clearances, the body was just 14 feet 6 inches (4.4 m) tall and 10 feet (3 m) wide, significantly lower and slightly narrower than most North American passenger types. And these constrained dimensions allowed Genesis diesels to work virtually anywhere on the North American passenger network.

Genesis's nonstandard appearance was the vision of Amtrak designer Cesar Vergara, who used an angular construction with flat surfaces, largely avoiding compound curves. Although many observers criticized the unusual Genesis design that contrasted sharply with classic North American diesels, and that didn't fulfill expected aesthetic appearances, still the Genesis was awarded accolades for industrial design. It has become the standard face of North American long-distance passenger railroading, leading many of Amtrak's named trains outside of the Northeast Corridor electrified zone.

↙ Among Amtrak's locomotives dressed in special liveries is P42 No. 42, painted to honor America's veterans. *Brian Solomon*

↗ Amtrak's 700 Series, P32AC-DMs, are specialized machines and the workhorses of *Empire Corridor* service between New York Penn Station and Albany. These "dual-mode" locomotives run as diesels using GE's twelve-cylinder 7FDL diesel engine and can draw electric current from line-side third rail using retractable third-rail shoes attached to the rear trucks. Amtrak 708 passes Stuyvesant, New York, on May 29, 2004. *Brian Solomon*

→ Amtrak's Genesis Series 1 were numbered in the 800 Series and employed GE's DASH 8–era technology. Some of these locomotives have survived in long-haul service on the *Auto Train* (a seventeen-hour run from Lorton, Virginia, to Sanford, Florida) because of their older braking controls that allow for superior handling characteristics, which is necessary for this exceptionally heavy passenger train. Amtrak 816 leads the southward *Auto Train* at Petersburg, Virginia. *Brian Solomon*

In 2011, Amtrak painted several of its Genesis diesels in heritage paint schemes to honor the railroad's fortieth birthday. Amtrak P42 156 wears the early 1970s' "bloody nose" livery. In December 2017, it leads the Boston section of the *Lake Shore Limited* along the Quaboag River at West Warren, Massachusetts. *Brian Solomon*

← Canadian intercity passenger operator VIA's first General Electric diesels were twenty-one P42s, ordered in 2001. On October 22, 2004, VIA Rail 908 leads a train of LRC tilting coaches out of Montreal's Central Station. VIA's paint livery was more sophisticated than Amtrak's various schemes. *Brian Solomon*

← ← In October 1997, Amtrak was making the transition from EMD F40s to GE Genesis on its long-distance services. P42 wears its as-delivered platinum mist with red, white, and blue striping—solid stripes all along the carbody. By contrast, the 800 Series P40s had featured an unusual faded stripe toward the rear of the locomotive. *Brian Solomon*

↙ ↙ Amtrak dual-mode P32AC-DM 710 leads New York City–bound *Empire Service* train 280 along the shores of the Hudson River at Breakneck Ridge near Cold Spring, New York. *Brian Solomon*

The three variations of Genesis differ primarily in their internal configurations. Amtrak's first orders were for the 800 Series engines, Genesis Series 1, incorporating DASH 8 technology, rated at 4,000 horsepower, designated as DASH 8-40BPs by GE but more commonly known as P40s on Amtrak. The first engines arrived from GE in 1993; ultimately, forty-four were built. Only a handful remain in Amtrak service in 2020, but surviving units work as preferred power for the Virginia–Florida *Auto Train*, in part because of their braking abilities, which allow engineers better control of the train's unusually long and heavy consist. Most numerous are the 4,200-horsepower Genesis P42DCs, the first of which entered service in 1996. These employ GE's DASH 9 components and are capable of 110 mph (177 km/h) operation. Amtrak bought 207 P42DCs (Nos. 1–207), with the last

delivered in 2001. In addition, Canada's VIA Rail bought twenty-one in 2001.

Most unusual are the P32AC-DMs, a specialized dual-mode diesel-electric/electric designed to operate on New York City–area third-rail electrified lines. These feature a smaller engine rated at 3,200 horsepower and three-phase alternating current traction motors; significantly, they can draw power from line-side third rail. Amtrak's P32AC-DMs are numbered in the 700 Series and are assigned to *Empire Corridor* Services between New York Penn Station and Albany, New York. Metro-North acquired a fleet of P32AC-DMs, some of which were paid for by the Connecticut Department of Transportation, for use on former New Haven Railroad branch-line services from Grand Central Terminal to Danbury and Waterbury, Connecticut.

Electro-Motive SD70MAC

Electro-Motive's SD70MAC was introduced in 1993 as the first mass-produced North American heavy-haul diesel-electric employing a modern three-phase AC propulsion system. This offered significant advantages over conventional direct current propulsion that had been employed by virtually all commercially produced North American diesel-electrics up until that time. Three-phase traction represented a revolutionary leap forward for the North American freight railroads. The innovation was made possible by advancing state-of-the-art European traction systems for use with the latest heavy diesel design. The innovation was made possible as the result of a significant financial commitment from Burlington Northern to purchase 350 locomotives for its coal-service fleet. BN's chairman, Gerald Grinstein, was famously quoted in *Trains* magazine stating that "[the SD70MAC] may very well represent the most dramatic step forward since diesel replaced steam."

Three-phase AC motors provide distinct advantages for locomotive traction, employing a simple design that delivers a greater output than DC motors of the same size while having an inherent ability to automatically correct for locomotive wheel slip; this combination provides superior adhesion and thus significantly greater pulling power. Historical complications with AC motor control were overcome through advances in high-voltage microprocessor controls that had been developed for high-speed electrics in the late 1970s, notably for Deutsche Bundesbahn (West German Federal Railways) Class 120 (see page 156). During the 1980s, further advances in three-phase traction aimed toward high-speed rail applications produced commercially successful trains in Germany, France, and Japan.

The successful application of three-phase traction in Germany encouraged Electro-Motive to partner with Siemens in adapting this technology for practical and reliable application on North American heavy-haul diesel locomotives. Electro-Motive experimented with a pair of four-axle prototypes in the early 1990s, designated F69PHACs. Later, it produced a quartet of three-phase AC, six-axle, six-motor demonstrators adapted from its recent SD60M, designated SD60MACs, and assigned to heavy coal service.

Once satisfied with the potential for the new design, BN placed its historic order for SD70MACs, which were assigned to Wyoming's Powder River Basin coal traffic. Three 4,000-horsepower three-phase traction locomotives were intended to perform the same work previously accomplished by five older 3,000-horsepower direct current traction diesels. BN successor BNSF placed repeat orders for the SD70MAC. Following BN/BNSF's successful application, Conrail and CSX also bought SD70MACs. Ultimately, three-phase AC traction became the new North American standard. In 2005, Electro-Motive ended production of the SD70MAC, introducing its more advanced SD70ACE.

→ Three new Burlington Northern SD70MACs work the windswept landscape of Wyoming's Powder River basin on May 28, 1995. Three-phase AC traction allowed BN significant cost savings in its movement of low-sulfur Powder River bituminous coal. *Brian Solomon*

↘ Classic 1990s heavy freight: a trio of matching Burlington Northern SD70MACs work, as intended, on an eastward loaded Powder River unit coal train near Edgemont, South Dakota. *Brian Solomon*

The sun shines on a heavy Powder River coal train as it ascends Nebraska's Crawford Hill on May 29, 1995. Although the train was down to walking speed because of the grade, its five SD70MACs—three at the front and two at the rear—kept the train moving without stalling. *Brian Solomon*

Bombardier TRAXX

The Bombardier TRAXX family is among the most common modern European locomotives, closely related to the latest generations of ALP models used in North America. TRAXX locomotives employ modular designs, incorporating common components and a common body shell tailored for high route availability across continental Europe. The TRAXX platform was derived from Germany's Class 101 electric built by ADtranz, a successful high-speed electric that entered production in 1996 and remains today as the standard electric used on Deutsche Bahn's InterCity and EuroCity long-distance passenger trains. Coincident with fundamental changes to European railway operations dictated by European Union legislation that have encouraged open-access operators and eased cross-border operation, the versatile and adaptable TRAXX models have emerged as some of the most widely used European locomotive types. The first TRAXX made its debut in 2000, and by 2015, more than 1,800 TRAXX locomotives were operating across sixteen European nations.

Historically, European variations in loading gauge, electrification, and signaling standards, combined with specific traffic demands and national desires to support domestic railway suppliers, resulted in myriad locomotive designs typically tailored to the specific needs of individual national railway systems. By contrast, Bombardier's TRAXX modular platform is aimed at overcoming national limitations; the model employs modern locomotive propulsion technologies, and with careful attention to clearance restrictions, models are capable of working over many lines across the European network.

The modern TRAXX carbody styles are designed to conform with more stringent EU crash-worthiness standards. Modern Bombardier's models include TRAXX AC for AC overhead electrification; TRAXX DC for

↙ The latest generation of TRAXX includes DB's Class 187 electrics, such as this one seen leading a southward freight along the Rhein at Sankt Goarshausen, Germany, in September 2019. The Class 187 uses the modular design of the TRAXX 3 platform, which features more elegant styling than the original TRAXX platforms. *Brian Solomon*

→ In April 2016, a pair of DB (German Railways) Class 185s lead an uphill freight on the old Gotthard Pass route near Zgraggen, Switzerland. Although many of the class predate the TRAXX classification, DB's Class 185 is now classified among the TRAXX dual-voltage F140AC types. It is similar to DB's Class 146. *Brian Solomon*

→ NJ Transit's ALP-46/
ALP-46A electrics are
part of Bombardier's
TRAXX family. On the
morning of December
11, 2015, NJ Transit
ALP-46A 4647 leads a
Hoboken-bound train at
Matawan, New Jersey.
Brian Solomon

← European Gateway
Services is an open
access freight provider
that connects the port
of Rotterdam in the
Netherlands with freight
hubs in central Europe.
One of its versatile
TRAXX electrics leads
a southward freight
along Germany's Rhein.
Brian Solomon

↙ Swiss operator
Berne–Lötchsberg–
Simplon Railway has
multi-voltage Class 185
TRAXX electrics for its
cross-border freight
operations. In September
2019, a northward BLS
container train glides
along the Rhein near
Boppard, Germany.
Brian Solomon

direct current overhead; and TRAXX MS, a multi-voltage type designed for cross-border services. In addition, these locomotives can be ordered with the "TRAXX Last Mile" option, a dual-mode technology wherein a small onboard diesel engine may be applied for slow speed switching beyond the reach of electrification. The TRAXX DE is a road diesel. Among TRAXX derivatives were 1,065-mm gauge electrics for the South African state-run railway Transnet.

In July 2018, Bombardier introduced its TRAXX 3 platform, initially with three basic variations. By that time, the builder had sold more than 2,200 locomotives in the TRAXX family. Among the most recent variations are a group of TRAXX MS3 models ordered in 2019 by Central European open-access passenger operator RegioJet.

In the United States, the ALP-46, built for New Jersey suburban rail service provider NJ Transit, was derived from the German Class 101 electric. It is a cousin to the TRAXX electrics built during the transition from ADtranz to Bombardier production. These high-horsepower electrics were built at Kassel, Germany. They use a standard B-B wheel arrangement with a body shell nearly 64 feet (19.5 m) long and just over 9 feet 8 inches (2.9 m) wide. Top design speed was 100 mph (160.9 km/h). In 2008, NJ Transit ordered a variation, thirty-six ALP-46As from Bombardier. Both ALP-46 and ALP-46A types work push-pull consists on NJ Transit suburban trains operating from Penn Station, New York, to various New Jersey towns.

Bombardier adapted its successful ALP-46A into a dual-mode (or "dual power") model ALP-45DP. This works as both a diesel-electric and as a straight electric, drawing current from overhead catenary, which enables passenger services originating on non-electrified lines to operate on electrified. Production included seventy-two locomotives, of which twenty were bought by Montreal-based suburban operator Agence Métropolitaine de Transport (AMT), and two orders totaling fifty-two units were bought by NJ Transit for services to New York's Penn Station and Hoboken Terminal in New Jersey.

Dual mode on the old Erie Railroad: working as a diesel, NJ Transit ALP-45DP 4526 accelerates away from Secaucus Transfer as it pushes a Hoboken Terminal–bound commuter train from Suffern, New York. In diesel mode, the ALP-45P is powered by a pair of Caterpillar twelve-cylinder 3512HD diesels. *Brian Solomon*

Siemens Vectron

Siemens's Vectron was introduced in 2010 as a versatile locomotive platform intended to support a variety of electric, diesel, and dual-mode models capable of cross-border European applications. This is Siemens's equivalent to Bombardier's TRAXX platform. Siemens Vectron electrics are specially tailored for a variety of applications on European railways and incorporate a host of the latest locomotive technologies. Vectrons employ state-of-the-art three-phase AC propulsion consists of Siemens Sibas 32 control systems with IGBT (insulated gate bi-polar transistor) semiconductor-type high-voltage inverters. These take electric current drawn from a variety of different overhead electrification standards and convert it into tractive power using modern three-phase AC-traction motors. To minimize unsprung weight, traction motors are directly mounted to trucks with rubber fittings. As with many modern electric locomotives, the Vectron features modern regenerative braking designed to feed 100 percent of braking power back into the electrical supply system via the overhead catenary, which makes for more efficient use of electric power.

Among Vectron variations at work in Europe are Finnish State Railways (known as the VR Group) Class Sr3. In 2015, Siemens sent a broad gauge Vectron prototype to Finland for testing and development of the Sr3 type for specific operating characteristics of Finnish railways that maintain service in difficult Arctic winter conditions.

↙ In 2015, Finnish Railways VR Group received a broad gauge Vectron from Siemens for testing in advance of production locomotives that were classed as Sr3s. The Sr3 Vectron is the most modern electric locomotive in Finland today. *Brian Solomon*

↗ In Germany, various open-access operators employ modern Class 193 Vectron electrics in freight service. BoxXpress Class 193-385 leads a long train of international containers at Lorch, Germany, in September 2019. *Brian Solomon*

→ MRCE is a European locomotive leasing company that supplies motive power to a variety of cross-border open-access operators. Its locomotives are equipped for multi-voltage operation and set up for numerous modern signaling systems, notably the European Train Control System. A sparkling clean MRCE Vectron works south along the Rhein in Germany. *Brian Solomon*

Siemens ACS-64

The ACS-64 is a distinct North American variation of the Vectron adapted for Amtrak's high-speed service on the Northeast Corridor between Boston and Washington, D.C., and associated connections. This draws current from three different overhead standards—12 kV at 25 Hz, and either 12.5 or 25 kV at 60 Hz. The locomotive is designed for low maintenance while providing high reliability. It employs rigorous North American crash-worthiness standards to ensure crew safety as mandated by the American Federal Railroad Administration.

Amtrak ordered seventy ACS-64 electrics (Nos. 600–669) from Siemens in October 2010 assembled at Siemens's Sacramento, California, plant. These are known as Amtrak's City Sprinters and are intended to haul up to eighteen Amfleet passenger cars at 125 mph

(201.2 km/h). Following extensive tests at the US Department of Transportation testing facility at Pueblo, Colorado, the ASC-64s were tested on the Northeast Corridor in May 2013, and the first new locomotive entered revenue service in February 2014. Amtrak's ACS-64 quickly supplanted its worn-out fleets of Swedish-designed AEM-7s and Alstom HHP8s on *Northeast Corridor Regional* and long-distance trains operating between Boston–New York–Philadelphia–Washington D.C., and Pennsylvania-sponsored *Keystone* trains between Philadelphia and Harrisburg. Philadelphia-based commuter operator SEPTA borrowed Amtrak ACS-64s for testing, ordering fifteen from Siemens to supplant its aging AEM-7/ALP-44 fleet on push-pull passenger sets. Like Amtrak's, these are numbered in the 600 Series.

↙ Amtrak's first ACS-64 600 entered revenue service on February 7, 2014, leading *Northeast Corridor Regional* train 171 westward from Boston to Washington, D.C. In this view, it races through Milford, Connecticut, on the former New Haven Railroad. *Brian Solomon*

↗ In 2019, Amtrak 606 was wrapped in a seasonal Coca-Cola advertising livery. Amtrak train No. 98 (*Silver Meteor*) is eastbound through Princeton Junction, New Jersey, on November 14, 2019, with Siemens-built ACS-64 606 that was wrapped for a temporary promotion to advertise Amtrak changing to supplying Coke products on board its trains. *Patrick Yough*

→ SEPTA train No. 6378 departs Woodbourne Station on June 14, 2019, for West Trenton, New Jersey, with ACS-64 905 shoving on the rear of a push-pull consist. *Patrick Yough*

Bibliography

Books

Allen, G. Freeman. *The Fastest Trains in the World.* London: Scribner, 1978.

Archer, Robert F. *A History of the Lehigh Valley Railroad—Route of the Black Diamond.* Berkeley, CA: Howell-North Books, 1977.

Binney, Marcus, and David Pearce, eds. *Railway Architecture.* London: Bloomsbury Books, 1979.

Bruce, Alfred W. *The Steam Locomotive in America.* New York: Bonanza Books, 1952.

Burgess, George H., and Miles C. Kennedy. *Centennial History of the Pennsylvania Railroad.* Philadelphia: Pennsylvania Railroad Company, 1949.

Bush, Donald J. *The Streamlined Decade.* New York: George Braziller, 1975.

Casey, Robert J., and W. A. S. Douglas. *The Lackawanna Story.* New York: McGraw-Hill, 1951.

Churella, Albert J. *From Steam to Diesel.* Princeton, NJ: Princeton University Press, 1998.

Collias, Joe G. *Mopac Power—Missouri Pacific Lines, Locomotives and Trains 1905–1955.* San Diego, CA: Howell-North Books, 1980.

Condit, Carl. *Port of New York.* Vols. 1 & 2. Chicago: University of Chicago Press, 1980, 1981.

Conrad, J. David. *The Steam Locomotive Directory of North America.* Vols. I & II. Polo, IL: Transportation Trails, 1988.

Cook, Richard J. *Super Power Steam Locomotives.* San Marino, CA: Golden West Books, 1966.

Corbin, Bernard G., and William F. Kerka. *Steam Locomotives of the Burlington Route.* Red Oak, IA: Bonanza Books, 1960.

Cupper, Dan. *Horseshoe Heritage: The Story of a Great Railroad Landmark.* Halifax, PA: Horseshoe Curve National Historic Landmark, 1996.

Dixon, Thomas W., Jr. *Chesapeake & Ohio—Superpower to Diesels.* Newton, NJ: Carstens, 1984.

Doherty, Timothy Scott, and Brian Solomon. *Conrail.* St. Paul, MN: MBI Publishing, 2004.

Dorsey, Edward Bates. *English and American Railroads Compared.* New York: Kessinger, 1887.

Drury, George H. *Guide to North American Steam Locomotives.* Waukesha, WI: Kalmbach, 1993, 2015.

———. *The Historical Guide to North American Railroads.* Waukesha, WI: Kalmbach, 1985, 1993.

———. *Santa Fe in the Mountains.* Waukesha, WI: Kalmbach, 1995.

Dubin, Arthur D. *More Classic Trains.* Milwaukee, WI: Kalmbach, 1974.

———. *Some Classic Trains.* Milwaukee, WI: Kalmbach, 1964.

Evans, Martin. *Pacific Steam: The British Pacific Locomotive.* Hemel Hempstead, UK: Model Aeronautical Press, 1961.

Farrington, S. Kip, Jr. *Railroading from the Rear End.* New York: Coward-McCann, 1946.

———. *Railroads at War.* New York: Coward-McCann, 1944.

———. *Railroads of Today.* New York: Coward-McCann, 1949.

Garmany, John B. *Southern Pacific Dieselization.* Edmonds, WA: Pacific Fast Mail, 1985.

Harlow, Alvin F. *The Road of the Century.* New York: Creative Age Press, 1947.

———. *Steelways of New England.* New York: Creative Age Press, 1946.

Harris, Ken, ed. *World Electric Locomotives.* London: Jane's, 1981.

Haut, F. J. G. *The History of the Electric Locomotive.* London: A. S. Barnes, 1969.

———. *The Pictorial History of Electric Locomotives.* Cranbury, NJ: Oak Tree, 1970.

Hayes, William Edward. *Iron Road to Empire—The History of the Rock Island Lines.* New York: Simmons-Boardman, 1953.

Heath, Erle. *Seventy-Five Years of Progress—Historical Sketch of the Southern Pacific.* San Francisco: Southern Pacific Bureau of News, 1945.

Hilton, George W. *American Narrow Gauge Railroads.* Stanford: Stanford University Press, 1990.

Hofsommer, Don L. *Southern Pacific 1901–1985.* College Station: Texas A&M, 1986.

Hollingsworth, Brian. *Modern Trains.* London: Arco, 1985.

Hollingsworth, Brian, and Arthur Cook. *Modern Locomotives.* London: Crescent Books, 1983.

Holton, James L. *The Reading Railroad: History of a Coal Age Empire.* Vols. I & II. Laurys Station, PA: Garrigues House, 1992.

Hungerford, Edward. *Men of Erie.* New York: Random House, 1946.

Joddard, Paul. *Raymond Loewy, Design Heroes.* London: Taplinger, 1992.

Johnson, Bob, with Joe Walsh and Mike Schafer. *The Art of the Streamliner.* New York: Metro Books, 2001.

Johnson, Howard, and Ken Harris. *Jane's Train Recognition Guide.* London: Collins, 2005.

Keilty, Edmund. *Interurbans without Wires.* Glendale, CA: Interurban Press, 1979.

Kiefer, P. W. *A Practical Evaluation of Railroad Motive Power.* New York: Steam Locomotive Research Institute, 1948.

Kirkland, John, F. *Dawn of the Diesel Age.* Pasadena, CA: Interurban Press, 1994.

———. *The Diesel Builders.* Vols. I, II, & III. Glendale, CA: Interurban Press, 1983.

Kirkman, Marshall M. *The Compound Locomotive.* New York: Kessinger, 1899.

Klein, Maury. *Union Pacific.* Vols. I & II. New York: Doubleday and Company, Inc., 1989.

Kratville, William, and Harold E. Ranks. *Motive Power of the Union Pacific.* Omaha, NE: Barnhart Press, 1958.

Lamb, W. Kaye. *History of the Canadian Pacific Railway.* New York: Macmillan, 1977.

LeMassena, Robert A. *Colorado's Mountain Railroads.* Golden, CO: Smoking Stack Press, 1963.

———. *Rio Grande to the Pacific.* Denver, CO: Sundance, 1974.

Loewy, Raymond. *The Locomotive (Its Esthetics).* New York: Universe, 1937.

Marre, Louis A. *Diesel Locomotives: The First 50 Years.* Waukesha, WI: Kalmbach, 1995.

Marre, Louis A., and Jerry A. Pinkepank. *The Contemporary Diesel Spotter's Guide.* Milwaukee: Kalmbach, 1985.

Marshall, John. *The Guinness Book of Rail Facts and Feats.* Enfield, UK: Guinness Superlatives, 1975.

Middleton, William D. *Landmarks on the Iron Road.* Bloomington, IN: Indiana University Press, 1999.

———. *When the Steam Railroads Electrified.* Milwaukee, WI: Kalmbach, 1974.

Morgan, David P. *Steam's Finest Hour.* Milwaukee, WI: Kalmbach, 1959.

Mulhearn, Daniel J., and John R. Taibi. *General Motors' F-Units.* New York: Quadrant Press, 1982.

Mullay, A. J. *Streamlined Steam, Britain's 1930s Luxury Expresses.* Devon, UK: David & Charles, 1994.

Nock, O. S. *British Locomotives of the 20th Century.* Vols. 2 & 3. London: Haynes, 1984.

———. *LNER Steam.* London: A. M. Kelley, 1969.

Overton, Richard C. *Burlington Route.* New York: Alfred A. Knopf, 1965.

Protheroe, Ernest. *The Railways of the World.* London: Routledge & Sons, n.d.

Ransome-Wallis, P. *World Railway Locomotives.* New York: Hawthorn, 1959.

Reck, Franklin M. *The Dilworth Story.* New York: McGraw Hill Book Co., 1954.

———. *On Time: The History of the Electro-Motive Division of General Motors.* Dearborn, MI: General Motors, 1948.

Rose, Joseph R. *American Wartime Transportation.* New York: Crowell, 1953.

Semmens, P. W. B. *High Speed in Japan.* Sheffield, UK: Platform 5, 1997.

Signor, John R. *Southern Pacific–Santa Fe Tehachapi.* San Marino, CA: Golden West Books, 1983.

Simmons, Jack. *Rail 150: The Stockton & Darlington Railway and What Followed.* London: Eyre Methuen, 1975.

Sinclair, Angus. *Development of the Locomotive Engine.* New York: MIT Press, 1970.

Snell, J. B. *Early Railways.* London: Octopus Books, 1972.

Solomon, Brian. *Alco Locomotives.* Minneapolis, MN: Voyageur, 2009.

———. *The American Steam Locomotive.* Osceola, WI: MBI Publishing, 1998.

———. *Amtrak.* St. Paul, MN: Voyageur, 2004.

———. *Brian Solomon's Railway Guide to Europe.* Waukesha, WI: Kalmbach, 2018.

———. *Bullet Trains.* Osceola, WI: MBI Publishing, 2001.

———. *Electro-Motive E-Units and F-Units.* Minneapolis, MN: Voyageur, 2011.

———. *EMD F-Unit Locomotives.* North Branch, MN: Specialty Press, 2005.

———. *EMD Locomotives.* St. Paul, MN: Voyageur, 2006.

———. *Railway Masterpieces: Celebrating the World's Greatest Trains, Stations and Feats of Engineering.* Osceola, WI: MBI Publishing, 2002.

———. *Southern Pacific Passenger Trains.* St. Paul, MN: Voyageur, 2005.

———. *Trains of the Old West.* New York: Michael Friedman Publishing Group, 1998.

Solomon, Brian, and Mike Schafer. *New York Central Railroad.* Osceola, WI: MBI Publishing, 1999.

Staufer, Alvin F. *Pennsy Power III.* Medina, OH: Alvin Staufer, 1993.

———. *Steam Power of the New York Central System.* Vol. 1. Medina, OH: Alvin Staufer, 1961.

Staufer, Alvin F., and Edward L. May. *New York Central's Later Power.* Medina, OH: Alvin Staufer, 1981.

Stilgoe, John R. *Metropolitan Corridor.* New Haven, CT: Yale Univiversity Press, 1983.

Stretton, Clement E. *The Development of the Locomotive.* London: Bracken Books, 1989.

Swengel, Frank M. *The American Steam Locomotive: Volume 1, Evolution.* Davenport, IA: Midwest Rail, 1967.

Taber, Thomas Townsend, and Thomas Taber Townsend III. *The Delaware, Lackawanna & Western Railroad, Part One.* Muncy, PA: Thomas T. Taber III, 1980.

Talbot, F. A. *Railway Wonders of the World.* Vols. 1 & 2. London: Cassell & Co., 1914.

Taylor, Arthur. *Hi-Tech Trains.* London: Apple Press, 1992.

Westwood, J. N. *Soviet Railways Today.* London: Citadel, 1963.

White, John H., Jr. *Early American Locomotives.* Toronto: Dover, 1979.

———. *A History of the American Locomotive—Its Development: 1830–1880.* Baltimore: Johns Hopkins Press, 1968.

Whitelegg, John, and Staffan Hultén. *High Speed Trains: Fast Tracks to the Future.* North Yorkshire, UK: Leading Edge Press, 1993.

Winchester, Clarence. *Railway Wonders of the World.* Vols. 1 & 2. London: Amalgamated Press, 1935.

Zimmermann, Karl R. *Erie Lackawanna East.* New York: Quadrant Press, 1975.

———. *The Remarkable GG1.* New York: Quadrant Press, 1977.

Brochures, Papers, and Manuals

Boston & Albany Railroad. *Facts about the Boston & Albany R.R.* 1933.

General Electric. *Dash 8 Locomotive Line.* n.d.

———. *Electro-Motive Division Model 567B Engine Maintenance Manual.* La Grange, IL: 1948.

———. *Electro-Motive Division F40PH-2C Operator's Manual.* La Grange, IL: 1988.

———. *Electro-Motive Division Model F3 Operating Manual No. 2308B.* La Grange, IL: 1948.

———. *Electro-Motive Division Model F7 Operating Manual No. 2310.* La Grange, IL: 1951.

———. *Electro-Motive Division Operating Manual No. 2300.* La Grange, IL: 1945.

———. *Electro-Motive Division SD70M Operator's Manual.* La Grange, IL: 1994.

———. *Genesis Series.* 1993.

———. *Electro-Motive Division SD80MAC Locomotive Operation Manual.* La Grange, IL: 1996.

———. *A New Generation for Increased Productivity.* Erie, PA: 1987.

United States Patents

809,974. January 16, 1906. W. R. McKeen Jr.

972,467. October 11, 1910. W. R. McKeen Jr.

972,502. October 25, 1910. E. H. Harriman and W. R. McKeen Jr.

973,622. October 25, 1910. E. G. Budd.

1,628,595. May 10, 1927. F. Kruckenberg et al.

1,631,269. June 7, 1927. P. Jaray.

1,727,070. September 3, 1929. F. Kruckenberg et al.

1,927,072. September 19, 1933. E. J. W. Ragsdale.

2,079,748. May 11, 1937. Martin Blomberg.

Reports and Unpublished Works

Chappell, Gordon. *Flanged Wheels on Steel Rails—Cars of Steamtown.* Unpublished manuscript.

Clemensen, A. Berle. "Historic Research Study: Steamtown National Historic Site Pennsylvania." Denver, CO: US Department of the Interior, 1988.

Warner, Paul T. "Compound Locomotives." Paper presented in New York, April 14, 1939.

———. *The Story of the Baldwin Locomotive Works.* Philadelphia, 1935.

Periodicals

Baldwin Locomotives. Philadelphia, PA (no longer published).

Diesel Era. Halifax, PA.

Jane's World Railways. London.

Journal of the Irish Railway Record Society. Dublin.

Locomotive & Railway Preservation. Waukesha, WI (no longer published).

Modern Railways. Surrey, UK.

Official Guide to the Railways. New York.

Rail. Peterborough, UK.

RailNews. Waukesha, WI (no longer published).

Railroad History, formerly *Railway and Locomotive Historical Society Bulletin.* Boston, MA.

Railway Age. Chicago and New York.

Railway Gazette, 1870–1908. New York (merged with *Railway Age* in 1927).

Today's Railways. Sheffield, UK.

Trains. Waukesha, WI.

Vintage Rails. Waukesha, WI (no longer published).

Index

Brimming with creative inspiration, how-to projects, and useful information to enrich your everyday life, Quarto Knows is a favorite destination for those pursuing their interests and passions. Visit our site and dig deeper with our books into your area of interest: Quarto Creates, Quarto Cooks, Quarto Homes, Quarto Lives, Quarto Drives, Quarto Explores, Quarto Gifts, or Quarto Kids.

© 2020 Quarto Publishing Group USA Inc.
Text © 2020 Brian Solomon

First Published in 2020 by Motorbooks, an imprint of The Quarto Group,
100 Cummings Center, Suite 265-D, Beverly, MA 01915, USA.
T (978) 282-9590 F (978) 283-2742 QuartoKnows.com

All rights reserved. No part of this book may be reproduced in any form without written permission of the copyright owners. All images in this book have been reproduced with the knowledge and prior consent of the artists concerned, and no responsibility is accepted by producer, publisher, or printer for any infringement of copyright or otherwise, arising from the contents of this publication. Every effort has been made to ensure that credits accurately comply with information supplied. We apologize for any inaccuracies that may have occurred and will resolve inaccurate or missing information in a subsequent reprinting of the book.

Motorbooks titles are also available at discount for retail, wholesale, promotional, and bulk purchase. For details, contact the Special Sales Manager by email at specialsales@quarto.com or by mail at The Quarto Group, Attn: Special Sales Manager, 100 Cummings Center, Suite 265-D, Beverly, MA 01915, USA.

24 23 22 21 20   1 2 3 4 5

ISBN: 978-0-7603-6810-7

Digital edition published in 2020
eISBN: 978-0-7603-6811-4

Library of Congress Cataloging-in-Publication Data

Names: Solomon, Brian, 1966- author.
Title: Rails around the world : two centuries of trains and locomotives / Brian Solomon.
Description: Beverly, MA : Motorbooks, imprint of The Quarto Group, 2020. | Includes bibliographical references and index. | Summary: "Rails Around the World is a visually glorious history depicting trains and locomotives at work in scenic locations throughout North America, Europe, and Asia"-- Provided by publisher.
Identifiers: LCCN 2020023862 (print) | LCCN 2020023863 (ebook) | ISBN 9780760368107 (hardcover) | ISBN 9780760368114 (ebook)
Subjects: LCSH: Locomotives--History--Pictorial works. | Locomotives--Pictorial works. | Railroad trains--History--Pictorial works. | Railroad trains--Pictorial works. | LCGFT: Illustrated works.
Classification: LCC TJ603 .S6685 2020 (print) | LCC TJ603 (ebook) | DDC 625.26--dc23
LC record available at https://lccn.loc.gov/2020023862
LC ebook record available at https://lccn.loc.gov/2020023863

Acquiring Editor: Dennis Pernu

Cover Image: Conway Scenic GP38 252, painted authentically for Maine Central, leads a passenger excursion eastbound over the Willey Brook Bridge at Crawford Notch, New Hampshire. The GP38 is a General Motors EMD diesel that is part of the family of locomotives including the famous F-units of the 1940s and 1950s, and the Dash-2 line of the 1970s and 1980s. *Brian Solomon*

Design: Elizabeth Van Itallie

Printed in China